CATASTROPHIC DISCLOSURE

THE DEEP STATE, ALIENS, AND THE TRUTH

KENT HECKENLIVELY, JD *AND* **MICHAEL MAZZOLA**

Post Hill PRESS

A POST HILL PRESS BOOK
ISBN: 979-8-89565-118-6
ISBN (eBook): 979-8-89565-119-3

Cover design by Jim Villaflores

This book, as well as any other Post Hill Press publications, may be purchased in bulk quantities at a special discounted rate. Contact orders@posthillpress.com for more information.

All people, locations, events, and situations are portrayed to the best of the author's memory. While all of the events described are true, many names and identifying details have been changed to protect the privacy of the people involved.

Post Hill
PRESS

Post Hill Press
New York • Nashville
posthillpress.com

Published in the United States of America

1 2 3 4 5 6 7 8 9 10

"The best way to keep a secret is to pretend to tell it."
—*Richard Dolan*, author and UFO researcher

CONTENTS

PREFACE

By Kent Heckenlively

I'm cowriting this book because a congressional hearing made me angry. I hate it when government officials lie to the public.

Although I've published twenty books on corruption in science, the COVID-19 task force, Big Tech companies, and the media, I've never written a single word about unidentified flying objects (UFOs), or (as they're known by their more current designation) unidentified anomalous phenomena (UAPs). I've never seen a UFO and have not had any "missing time" experiences. I've had one or two people tell me their UFO experiences, but nothing more than the average person has probably heard from friends or family. I do have a recreational interest in UFOs. I'm curious when there's a new documentary on the subject, and I've probably read about fifteen books on UFOs.

For the past nineteen years, I've been paid by the State of California to teach science. But I was trained as an attorney (passed the California bar in 1990), which taught me to weigh evidence and credibility in legal cases and people's stories.

The congressional hearing on UFOs/UAPs on July 26, 2023 convinced me this was a worthy subject for a book. Somebody was lying; I just didn't know who: the government, or the whistleblowers. This is from a July 28, 2023, CBS News article about the hearing:

> David Grusch, who served for 14 years as an intelligence officer in the Air Force and National Geospatial Intelligence Agency, appeared before the House Oversight Committee's national security subcommittee alongside two former fighter pilots who had firsthand experience with UAPs.
>
> Grusch served as a representative on two Pentagon task forces investigating UAPs until earlier this year. He told lawmakers

that he was informed of "a multi-decade UAP crash retrieval and reverse-engineering program" during the course of his work examining classified programs. He said he was denied access to those programs when he requested it and accused the military of misappropriating funds to shield those operations from congressional oversight. He later said he had interviewed officials who had direct knowledge of aircraft with "nonhuman" origins, and that so-called "biologics" were recovered from some craft.[1]

The first thing an attorney would do upon hearing an account such as this would be to ask, "What are the possible truths?" He would be hesitant to claim to know the truth after hearing only one side. Perhaps the person speaking is honest; perhaps he's a liar.

More often than not, you find that the truth lies somewhere in the middle. Both sides can have a good-faith belief they are right but have gotten some facts wrong. And in order to resolve the complaint, you need to navigate between the two extremes.

You must be in control of the facts but also understand that much of what you believe may not rest on solid ground. You should be curious but remain skeptical.

Let's consider the possibilities with the above excerpt about David Grusch.

The first is that he is telling the truth, and that the US military has been engaging in crash retrieval of UFOs for decades, reverse engineering them in hidden programs, and hiding this research from congressional oversight as well as the taxpaying public, and is in possession of alien bodies.

The second is that Grusch is lying, and the US military has not been engaging in crash retrieval of UFOs for decades, there are no programs to reverse engineer them, there has been no concealment of research from Congress, and the government holds no alien bodies.

The third possibility is that Grusch is telling some truths and some lies. Whether he is knowingly telling some lies or is genuinely mistaken is another question which we will also examine.

When trying to understand such complicated claims, we must ask the necessary questions, seek the answers without fear, and yet always be willing to revise our opinions in light of new evidence.

▲   ▲   ▲

My goal is to provide you with the documentation that I believe supports the narrative I am proposing. I want you to see the evidence upon which I'm basing my opinion, and you can determine whether I am drawing reasonable conclusions. With that in mind, I want to go to several sections of the July 26, 2023 congressional hearing on UFOs/UAPs. Congressman Glenn Grothman (R-Wisconsin) had this to say in his opening statement:

> The lack of transparency regarding UAPs has fueled wild speculation and debate for decades, eroding public trust in the very institutions that are meant to serve and protect them, as is evidenced by the large number of people we have here. I also want to point out, in 1966, President Gerald Ford claimed to have seen a UFO. And in 1969, in Georgia, Jimmy Carter claimed to have seen a UFO. So this has led Congress to establish entities to examine UAPs. The National Defense Authorization Act of 2022 established the All-Domain Anomaly Resolution Office, or AARO, to conduct or coordinate efforts across the Department of Defense and other federal agencies to detect, identify, and investigate UAPs.[2]

The congressman went on to note that AARO's budget was classified, and he believed that Americans had been lied to on the subject for more than fifty years.

The opening statement of Congresswoman Anna Paulina Luna (R-Florida) also seemed to suggest that the hearing was an honest attempt to get to the truth:

> The circumstances surrounding UAPs has [sic] captivated the intention [sic] of the American people for decades, ingrained even in the minds of our nation's leaders from Jimmy Carter to Barack Obama, Hillary Clinton to Donald Trump, Marco Rubio to Chuck Schumer, John Radcliffe to National Security officials. Yet, from Roswell, New Mexico to the coast of Jacksonville, Florida, the sightings of UAPs have rarely been explained by the people who have firsthand accounts of these situations. This is largely due to the lack of transparency by our

own government and the failure of our elected leaders to make good on their promises to release explanations and footage and mountains of overclassified documents that continue to be hidden from the American people.[3]

Congresswoman Luna's statement hit the main points, mentioning the Roswell incident of 1947 and ending with the revelations about the sightings off the East Coast of the United States. We'll discuss these more later.

The first witness was Ryan Graves, a former F-18 pilot with more than a decade of service, including being deployed during Operation Enduring Freedom and Operation Inherent Resolve. In 2014, he was stationed at Naval Air Station Oceana in Virginia Beach, Virginia, as an F-18 Foxtrot pilot. In his testimony, he said:

> After upgrades were made to our jet's radar system, we began detecting unknown objects operating in our airspace. At first, we assumed they were radar errors, but soon we began to correlate the radar tracks with multiple onboard sensors, including infrared systems, eventually through visual ID. During a training mission in warning area W72, ten miles off the coast of Virginia Beach, two F-18 Super Hornets were split by a UAP. The object, described as a dark gray or a black cube inside of a clear sphere, came within fifty feet of the lead aircraft, and was estimated to be five to fifteen feet in diameter....
>
> Sometimes these reports are reoccurring with numerous recent sightings north of Hawaii and the North Atlantic. Other veterans are also coming forward to us regarding UAP encounters in our airspaces and oceans. The most compelling involved observations of UAP by multiple witnesses and sensor systems. I believe these accounts are only scratching the surface and more will share their experiences once it is safe to do so. In closing, I recognize the skepticism surrounding this topic. If everyone could see the sensor and video data I witnessed, our national conversation would change.[4]

The average citizen, upon hearing this account, could only conclude that something unexplained was happening in our skies. A Navy pilot who has seen convincing video and sensor evidence of UFOs? And multiple other witnesses have observed UAPs? According to Graves' testimony, encounters with these objects were so common that noting their continued presence was part of regular preflight briefs.

The next witness set off my bullshit detector. He was just too happy, too bouncy, too convinced that nothing he said would derail his career or subject him to ridicule from family, friends, or colleagues. I'll let him speak for himself:

> Mr. Chairman, ranking members and congressmen, thank you. I'm happy to be here. This is an important issue, and I'm grateful for your time. My name is Charles David Grusch. I was an intelligence officer for fourteen years, both in the US Air Force, both active-duty Air National Guard and Reserve at the rank of major. Most recently, from 2021 to 2023, at the National Geospatial Intelligence Agency, NGA, at the GS-15 civilian level, which is the military equivalent of a full-bird colonel.

> I was my agency's co-lead in unidentified anomalous phenomena and trans-medium object analysis, as well as reporting to the UAP Task Force, UAPTF. Eventually, once it was established, the All-Domain Anomaly Resolution Office, AARO. I became a whistleblower through a PPD-19 Urgent Concern filing in May 2022 with the intelligence community inspector general. Following concerning reports from multiple esteemed and credentialed current and former military and intelligence community individuals, that the US government is operating with secrecy above congressional oversight with regards to UAPs. My testimony is based on information I've been given by individuals with a longstanding track record of legitimacy and service to this country...

> I am driven by a commitment of [sic] both the truth and transparency, rooted in our inherent duty to uphold the United States Constitution and protect the American people. I'm asking Congress to hold our government to this standard and thoroughly investigate these claims, but as I stand here under oath now, *I am speaking to the facts as I've been told them*[5] [emphasis added].

Grusch's testimony seems too "on the nose"—meaning so predictable as to either be a trap or lack imagination. In our modern world, who says, "I am driven by a commitment of both[…]truth and transparency, rooted in our inherent duty to uphold the United States Constitution and protect the American people?"

Make no mistake, I *want* to believe that sentiment. But I suspect there is something going on, such as when he says, "I am speaking to the facts as [I've] been told them." That's a curious construction of words, almost as if he's preparing his retreat for when somebody points out his lies and deceptions. Further in Grusch's testimony was this exchange with Congressman Robert Garcia (D-California):

> **CONGRESSMAN GARCIA:** My last question, and I know you have also said some of these answers in the past, but we're trying to get them on the public record, as well, which is really important. Mr. Grusch, finally, do you believe that our government is in possession of UAPs?

> **DAVID GRUSCH:** Absolutely, based on interviewing over forty witnesses over four years.[6]

Grusch goes on to say he also knows the location of these crashed alien vehicles.

Consider me suspicious when somebody claims in public to know where crashed UFOs are located and yet, in a congressional hearing, does not disclose where they are located. Then there was this exchange between Congresswoman Luna and Grusch:

> **CONGRESSWOMAN LUNA:** And then, my last question for you, before I move to Mr. Graves, is: you received prior approval from the Defense Department [DOD] to speak on certain issues, correct?

> **DAVID GRUSCH:** Correct. Through DOPSR, [which is the] DOD Prepublication and Security Review [department]. And I just want to remind the public, they're looking at it from a security perspective. These are my own personal views and opinions, not the department's.

> **CONGRESSWOMAN LUNA:** Okay. I'm asking that, though, mainly because I think that there are many people that would like to dis-

credit you. So, it does bring a certain amount of credibility to your testimony.

DAVID GRUSCH: I'm required by law to do that as a former intelligence officer. Or I go to jail for revealing classified information.[7]

This continues my trouble with Grusch's testimony. He's a former intelligence officer, and he's discussing this information with the *approval* of the intelligence agencies he worked for.

In my mind, there are three possibilities that might explain why an intelligence officer would release to the public what would seem like highly classified information.

First, the intelligence agencies are genuinely confused as to what is happening in our skies and oceans and believe the time has come to release this information to the public.

Second, nothing like what Grusch is describing is true, and this is all some type of psychological warfare by the intelligence agencies to determine the gullibility of the public.

Third, the intelligence agencies are telling the truth about crashed alien vehicles, but they will use that information in a way to financially enrich themselves, or for other nefarious means.

I wish I knew which possibility was most likely to be true.

The testimony of David Fravor, who in 2004, as the commanding officer of the US Navy's Strike Fighter Squadron 41 (also known as "Black Aces"), was part of an incident off the coast of San Diego. This was part of his testimony about what happened after he launched his plane from the USS *Nimitz*:

> As we proceeded to the west, the air controller was counting down the range to an object that we were going to, and we were unaware of what we were going to see, when we arrived. The controller had told us that these objects had been observed for over two weeks coming down from over 80,000 feet, rapidly descending to 20,000 feet, hanging out for hours, and then going straight back up. For those who don't realize, above 80,000 feet is space.

We arrived at the location at approximately 20,000 feet in the controller called Merge Plot, which means that our radar blip was now in the same resolution cell as the contact. As we looked around, we noticed that we saw some whitewater off our right side. It's important to note that the weather on this day was as close to perfect as you could ask for off the coast of San Diego: clear skies, light winds, calm seas, no whitecaps from waves, so, the whitewater stood out in a large blue ocean. All four of us, because we were in F-18s, we had pilots and Wizzo [a Weapons Systems Officer, or WSO, who rides in the back of the jet and monitors radar targets] in the back seat, looked down and saw a white Tic Tac object with a longitudinal axis pointing north-south and moving very abruptly over the water like a ping-pong ball. There were no rotors, no rotor wash, or any visible control surfaces like wings.[8]

David Fravor seemed like a genuine witness to me, describing what he'd seen, offering his opinion that these were unlikely to be manmade craft, but also open to the possibility that they might be our own advanced weapons.

But if the latter were true, why were those who were controlling them showing them off in such a manner to our own military? Could it be to convince them that if it came to a fight, those controlling these secret programs could easily defeat our public military forces?

The following exchange seemed to express the purpose of this public event.

CONGRESSMAN GROTHMAN: Okay, this is for any one of you. Based off each of your experiences, *do you believe UAPs pose a potential threat to our national security?* [Emphasis added.]

DAVID FRAVOR: I'll do it. Yes, and here's why. The technology that we faced was far superior to anything we had. And you could put that anywhere. If you had one, you captured one, you reverse engineered it. You get it to work, you're talking something that can go into space, go someplace, drop down in a matter of seconds, do whatever it wants and leave. And there's nothing we can do about it. Nothing.[9]

As I considered Fravor's testimony, one scenario I could imagine is that a secret group of individuals was in possession of this technology, and they were using it in a manner to best benefit themselves. If a cabal of military contractors and their agents in the government had this technology, wouldn't it be a great strategy to harass our military assets, spurring congressional demands for massive spending to defend against this threat? You spent taxpayer dollars to reverse engineer recovered alien technology, then you get even more money to defend against the threat.

It would be an extremely well-thought-out criminal plan.

Again, I found myself conflicted about David Grusch's previous testimony, as he seems to confirm some of my suspicions. And yet he strikes me as somebody who's lying.

As a lawyer, I'm aware that the best way to get somebody to believe a lie is to tell that person something they want to believe. Like the character Fox Mulder in the long-running television show *The X-Files* says about UFOs and other paranormal phenomena, "I want to believe." But I've experienced enough in life to realize that sentiment should be more along the lines of, "I want to believe what's true, and yet I know the world is filled with liars."

This is more from Grusch's congressional testimony, in an exchange with Congressman Jacob Moskowitz (D-Florida):

CONGRESSMAN MOSKOWITZ: Do you have knowledge, or do you have reason to believe that there are programs in the advanced tech space that are unsanctioned?

DAVID GRUSCH: Yes, I do.

CONGRESSMAN MOSKOWITZ: And when you say that they're above congressional oversight, what do you mean?

DAVID GRUSCH: Complicated question. So, there is some, I would call it abuse, here. So, congressional oversight of conventional Special Access Programs [SAPs], and I'll use Title 10. So, DOD, as an example, right? So, 10 US code, section 119, discusses congressional oversight of SAPs, discusses the defense secretary's ability to waive congressional reporting. However, the Gang of Eight [a bipartisan group of eight US senators whom the executive branch briefs on intelligence matters], is at least supposed to be notified [sic] a waived or unacknowledged SAP is created. And that's public law—

CONGRESSMAN MOSKOWITZ: I mean, I don't want to cut you off, but how does a program like that get funded?

DAVID GRUSCH: I will give you generalities. I can get very specific in a closed session, but a misappropriation of funds and self-funding.[10]

So, Grusch testified in the United States Congress about the suspicions many people have had for decades about our government and the UFO issue. Either officials at very high levels had been lying for years about it, or, for some reason I couldn't understand, this former intelligence official was now lying in the halls of Congress.

I wanted answers.

▲   ▲   ▲

By coincidence, I had a source close to one of the major figures of the study of UFOs, a documentary filmmaker named Michael Mazzola. Michael had optioned and filmed a documentary of my 2020 *The New York Times* bestselling book with Judy Mikovits, *Plague of Corruption: Restoring Faith in the Promise of Science*.

Prior to working with me, Michael had developed a close friendship with former emergency room physician Steven Greer, after Michael most notably directed the well-known 2017 UFO documentary *Unacknowledged*, about Greer's nearly three-decade investigation of the phenomenon.

I'm always intrigued by the personality of public figures, and over the years I've peppered Michael with questions about Greer and the UFO field. Michael has often told me I should collaborate with Greer on a book about the subject, but I'd always been too busy to agree to such a project.

The congressional hearing changed all that.

No matter what else was going on in my life, no matter how many books I'd promised to other people, this story had to move to the head of the line.

I called Michael and told him I was ready to write the UFO story.

"That's great," he said, then proceeded to tell me something very troubling. "We're getting multiple reports from whistleblowers of the use of an expression we've never heard before. Sources in the government are telling these whistleblowers that the government is engaging in 'controlled disclosure' about aliens, because they want to avoid 'catastrophic disclosure' about what's really been happening."

"Which is what?" I asked.

"The alien threat narrative," he said. "They're going to twist the facts to make the aliens a danger to humanity. The reality is that at any time in the past seventy-five years, they could have destroyed our civilization. They're not here to conquer us. The reality is, we can detect them with our sensors; we do so almost every day. We have technology which can bring down these crafts, and...we have done [that]. The aliens have exhibited an almost infinite patience with us, even when we experiment on the aliens we've recovered from these crashes, or created false narratives in which the aliens are abducting and experimenting on humans. The 'catastrophic disclosure' is not that the aliens are monsters, but that human beings in the military and intelligence services are the real monsters."

Yes, I understood. I was creating a counternarrative to what was presented in those congressional hearings.

"I want to talk to Greer," I said.

"I'll make it happen," Michael replied.

$$\blacktriangle \quad \blacktriangle \quad \blacktriangle$$

When I first spoke to Greer, he thought my proposed book should tell the arc of his Disclosure Project, started in the early 1990s, with an emphasis on evidence that had previously been kept secret. The story would cover the mostly successful effort he's undertaken over the years to convince top officials at the Pentagon and in Congress that the secret Special Access Programs that Grusch mentioned are unconstitutional because they lack congressional oversight. As a result of these programs' being illegal and unconstitutional, any secrecy agreements signed by individuals who participate in these programs are not enforceable.

Simply stated, if government money is used in any of these projects, there *must* be congressional oversight. If there has been no congressional oversight, these programs exist outside of our civilian government and are by definition unconstitutional, so it is not a betrayal of national security to discuss them.

Greer did not mince words when he said he considered these programs to be not only illegal but "treasonous," and that they have cost some people their lives.

In Greer's opinion, rogue elements of our government have waged a decades-long covert aggressive military campaign against these extraterrestrial visitors. But the recent congressional hearings herald a dangerous new escalation, an overt war against these visitors, which might lead to a cataclysm for our planet.

Greer is in favor of disclosure but worries that current efforts are designed to serve a militaristic totalitarian takeover, fueled by fear of these aliens.

I suggested to Greer that we collaborate. I told him I'd write a few chapters, then he could tell me what he thought. He agreed, and we ended our first conversation.

▲　▲　▲

While the tone of our first conversation had been friendly, our second conversation got off to a rocky start.

The substance of his remarks was that despite the books I'd written, those issues were nothing compared to the complexity of the UFO story.

I asked how Greer would feel if I collaborated instead with Michael Mazzola, our mutual friend and Greer's longtime filmmaking partner.

Greer thought that was an excellent idea. He spent some more time going over the dangers of this issue, then said, "And I'll give you one last warning: a little knowledge on this is a dangerous thing. Because people think they can go on Google, smush it around, [and] that they'll be able to come up with an informed opinion. They'll think they know what they're talking about, but they don't."[11]

He also told a story I'd heard him tell several times before in interviews. This is how he described the series of events from an interview he gave on February 20, 2023, on *The Shawn Ryan Show*. According to Greer, there were "friendlies" in the US government, who reached out to him because of his work documenting alien encounters.

By 1993 I was being asked to go up and brief the director of the CIA for Bill Clinton, R. James Woolsey. I'm a thirtysomething medical doctor. The letter that came to me, and this is the key point I want to make, stated, and I have it in my archives, it's going to come out publicly. It's says, 'You're going to be the first person' to brief the Clinton administration and the director of the CIA on this issue.

Because they'd made inquiries and had been denied access. Huge red flags should go up over everybody's head now. You're talking about the president of the United States, the director of the CIA, making an inquiry, through official channels about something they know is happening, and they're being denied access.[12]

On the subject of David Grusch, Greer was skeptical. He believed Grush was well-intentioned but had been led astray by intelligence assets like Luis Elizondo and Colonel John Alexander.

Greer said that even decorated military leaders can fall victim to the disinformation. "They were actually able to gaslight poor President Eisenhower," he said. "When this whole thing went sideways, it was because Eisenhower lost control of it, and it was not his fault. He just trusted people around him, who were devils who betrayed him. I know this for a fact, because I dealt with some people who were still alive when I was first doing this, in the early nineties, who were friends of Eisenhower. They actually worked in the White House with him as younger guys."[13]

When the interview was over, I called Michael and asked if he'd be my coauthor.

He quickly agreed.

Michael and I were now a team.

And hopefully, with my experience as an investigative writer, and Michael's as a documentary filmmaker, we could finally get to the bottom of the UFO mystery.

⋏　⋏　⋏

On August 20, 2024, William Morrow published Luis Elizondo's book *Imminent: Inside the Pentagon's Hunt for UFOs*, which noted Elizondo's position as "Former Head of the Department of Defense's Advanced Aerospace Threat Identification Program." For the week of August 25, 2024, it ranked second on Amazon's nonfiction list. I bought a copy and quickly read through it to see if I might share some of Greer's skepticism about Elizondo.

I barely made it out of his short (two-page) introduction when I ran across a paragraph that set off my alarm bells.

> While there are valid reasons for secrecy around some aspects of UAP, I do not think humanity should be kept in the dark about the fundamental fact that we are not the only intelligent life in the universe. The United States government and other major governments have decided its citizens do not have a right to know, but I could not disagree more.[14]

When I read this, I imagined a witness in front of me in a courtroom, with the bailiff asking them to raise their hand and promise to tell "the truth, the whole truth, and nothing but the truth." I pictured Luis Elizondo in that witness chair, and he didn't say, "I do." Instead, he replied that there were some "valid reasons" he will lie (or not tell you the truth), but that what he does say will be true.

This scenario does not fill us with confidence. Maybe it's the difference in our backgrounds. We're not a part of the world of secrets and shadows. Elizondo freely admits that he is. Perhaps he's trying to be honorable, but then so are we. They say a man is known by his friends, so who does Elizondo count among that group?

One of the most famous (or infamous) of Elizondo's friends is James Clapper, known to many as "America's Most Notorious Liar." In his book, Elizondo brags about having been in a meeting with Clapper when Clapper got the call from President Barack Obama to be the next director of national intelligence, and that he considered the appointment of Clapper a "big win for America."[15]

Elizondo certainly establishes himself as being in the center ring of power of the American intelligence community in 2010 when Clapper became the director of national intelligence.

But that may not necessarily be a source of pride, because, a few years later, Clapper would tell one of the greatest lies ever uttered by a government official—that the intelligence agencies were not spying on otherwise law-abiding American citizens. This is from a July 2013 National Public Radio story about Clapper's big lie:

> After telling Congress that the National Security Agency does not collect data on millions of Americans, National Intelligence Director James Clapper has issued an apology, telling Senate Intelligence Committee Chairwoman Dianne Feinstein that his statement was "clearly erroneous."
>
> Secret documents leaked by former NSA contract worker Edward Snowden have shown that the agency has been collecting metadata from phone records on millions of Americans. The documents have indicated an ability to conduct surveillance on Americans' internet activities.[16]

One might conclude that Clapper is the deep state's designated liar, which explains his current job as a political commentator on CNN.

But Clapper isn't the only controversial figure whom Elizondo counted as a friend. There was also General James "Mad Dog" Mattis, the secretary of defense for the first Trump administration. Elizondo described Mattis as a "true warrior monk," said that he (Elizondo) served as "the first-ever counterintelligence coordinating authority in Afghanistan," and said that Elizondo gave Mattis the information on which he based his decisions to "save lives or annihilate them."[17]

This above passage concerns us. Elizondo starts out by praising Mattis in heroic terms (putting the subject at ease), then slips in that he was the guy giving Mattis the information upon which he decided whom to kill.

Elizondo is a persuasive communicator, but we're not sure we trust him. He follows up with this account of collaborating with Mattis to thwart a rocket attack on their base in Afghanistan:

> One of my early bonding moments with him was when I came to him with what is known as immediate flash traffic. "Sir, there's going to be a rocket attack against us in like ten minutes, right here on the airfield," I said. "We've got to get some support."
>
> He turned to one of his underlings and started yelling. "Get all the helos in the air *now*! Gunners on all perimeters. I want these

f---ers dead!" When the coordinates had been relayed, he turned to me and said with a chuckle, "I hope you're right, Lue."[18]

It's a perfect little story. A few paragraphs and you feel that not only have you been on a base in Afghanistan just before a rocket attack, but you've bonded with the commander.

Elizondo and Mattis were genuinely "brothers-in-arms."

And then there's the matter of the "Mad Dog" nickname. Did Mattis deserve it?

Well, when Mattis became secretary of defense under President Trump, this was one of his ideas, as reported by *The New York Times* national security correspondent David E. Sanger in his 2018 book *The Perfect Weapon: War, Sabotage and Fear in the Cyber Age.*

> A year into Donald J. Trump's presidency, his defense secretary, Jim Mattis, sent the new commander-in-chief a startling recommendation: with nations around the world threatening to use cyberweapons to bring down America's power grids, cellphone networks, and water supplies, Trump should declare he was ready to take extraordinary steps to protect the country. If any nation hit America's critical infrastructure with a devastating strike, even a non-nuclear one, it should be forewarned that the United States might reach for a nuclear weapon in response.[19]

Perhaps the "Mad Dog" moniker was appropriate. But maybe I'm making too much of Elizondo's associations. What about the actions of the man himself? Is there anything there that might prompt suspicion?

Again, we return to Elizondo's own book, in which he admits, "I was tasked with running certain elements of Guantanamo Bay and the secret prison there known as Camp 7."[20] And he goes into detail about his later notoriety.

> My work with Guantanamo Bay brought endless rounds of drama and stress. An attorney for one of the 9/11 suspects labeled me in open court as the "US Czar of Torture." From that moment, I would be forever branded by some as the nation's Darth Vader. At one point I was informed that Europe had issued an open arrest warrant for me and anyone involved

in the notorious Rendition, Detention, and Interrogation (RDI) program of high-value detainees. (HVDs).[21]

It seems to us there are two possibilities with Elizondo. The first is that he is exactly the type of person the intelligence agencies would use to lie to us. The second is that he is exactly the type of person who would know what was true and would be in the best position to credibly tell us.

We wish we could tell you which is more likely. We cannot.

With that caveat, let's consider some of the amazing claims Elizondo makes.

One of the earliest revelations in the book is a document referred to as the Wilson/Davis memo, after the two individuals involved, Vice Admiral Thomas R. Wilson, who was director of the Defense Intelligence Agency from 1999 to 2002, and Eric Davis, an astrophysicist with high-level national security clearance. Davis had allegedly informed Wilson about the secret programs that were being funded through specific black program (budget items not revealed to congress) line items in the budgets that he reviewed.

Wilson later memorialized the results of his investigation in a thirteen-page summary, which was kept in his safe and not released until after he died. As Elizondo wrote:

> The Wilson/Davis memo created a sensation for good reason. Vice Admiral Wilson, curious about certain black program line items in the budgets that crossed his desk, began asking questions. He got a meeting with representatives of a certain aerospace corporation—and their attorney. Wilson put his cards on the table: Just what were they doing with this specific line item?
>
> He discovered that the contractor was part of an extremely secret program focused on retrieving and reverse engineering crashed advanced vehicles of unknown origin and not made by humans. I learned the larger program is referred to as the Legacy Program and involves various elements of the US government and US defense contractors. The contractors took the wreckage into their possession, and the security enveloping these projects was beyond top secret. In fact, the contractor's attorney brashly informed the admiral that if he continued

to ask questions, it could get him fired and cause him to lose his pension.[22]

We find these claims to be consistent with what Greer and many others in the field, including the congressional whistleblowers, report.

Something is happening behind the scenes: the government is spending a lot of money on it, and the government and others don't want us to know about it.

But if Greer is to be believed, if it's disinformation from the intelligence agencies, it's got just enough truth mixed to make us believe before they drop the lie. As we read the book, that's what we were anticipating.

One of the observations Elizondo makes is that these alien visitors seem to be drawn to water and to nukes, particularly nuclear warships.

> The Nimitz and the Roosevelt sightings happened on the open sea. In the Belgian Congo in 1952, the UAP fled the uranium mines and escaped in the direction of Lake Tanganyika, the second largest freshwater lake in the world. And in that 1988 UAP incident on Lake Erie, as the UAP descended, the Coast Guard investigators observed "that the ice was cracking and moving abnormal amounts as the object came closer." I thought of the Tic Tac darting around a roiling, bubbling circle of the Pacific Ocean in 2004. Maybe, when the water or ice is agitated, these ships can more easily strip off and harvest the hydrogen atoms?[23]

While this is speculation, we found it to be one of the most interesting parts of the book. Elizondo details his conversations with the physicist Harold (Hal) Puthoff on the five commonalities associated with these craft. Puthoff says they are instantaneous acceleration, hypersonic velocity, low observability, trans medium travel (the ability to operate equally well in air and water), and antigravity.[24] Puthoff concludes that all of these traits could be explained by the creation of a time and space "bubble" around the craft.

> Hal explained that it turns out "if we had the right technology, we could warp space and time in a localized area, creating a localized 'bubble' around the craft."

Inside the bubble, one would experience space and time differently than someone outside the bubble ... like a diving bell, which protects a diver from the crushing depths around them.

How is this bubble created?

"In theory, there are only two ways to warp space-time: a lot of mass, or a heck of a lot of energy." A smile crept to his face. He added, "An obscene amount of energy."[25]

This is a big idea, and we don't claim to have the physics background to know whether it's plausible. But the explanation solves many of the vexing UFO mysteries, such as the impossible high-speed maneuvers that we know would kill human beings, the fuzzy pictures often taken of UFOs (the time and space bubble causes light distortion), as well as the possibility that the craft would emit some form of radiation, which up close could significantly damage people.

The other interesting part of this time and space bubble is that the perfect shape for such a craft would be a saucer, positioned in the center of this bubble. When Elizondo and Puthoff considered the other types of UFOs commonly observed, triangles and cigar-shaped craft, they realized that the addition of one, two, or even three more bubbles (created by a propulsion unit) could easily accommodate such craft.[26] On the question of the intentions of these aliens, Elizondo, approaches the question with a commendable humility. These are the three scenarios he presents:

1. The visitors are benevolent, and don't want to interfere with our existence. They just want to continue using Earth as a galactic way station for natural resources. Or, possibly, they're so benevolent that they hope to save us from ourselves.

2. They're malevolent; they are here to take from us and will show up in vast numbers in the future.

3. They're neutral; like humans they can do both good and bad, and they hope to observe and learn from us.[27]

It's to Elizondo's credit that he struggles with the most likely scenario. The visitors' actions seem benign for the most part, but Elizondo is suspicious of the abduction phenomenon and the interest in nuclear sites and capabilities.

In many ways it seems to fit a military concept known as initial preparation of the battlefield, but it seems to be an extremely long preparation.

Although Greer disbelieves Elizondo, the question he asks is the same many have wondered about: if the visitors are real, what are their intentions?

As far as we can determine, Elizondo thinks it most likely that these visitors are indifferent or mildly benevolent. But if history is any guide, it's usually not a good idea to trust those who arrive with advanced technology and promises of friendship.

CHAPTER ONE

THE TRINITY OF AMERICAN POWER: NUKES, UFOS, AND HOLLYWOOD

Michael thought that the first thing readers needed to understand was the paranoid nature of American leaders when the first UFOs were recovered in the 1940s. He suggested I review the famous essay "The American Century" by *Life* magazine publisher Henry Luce, published in early 1941, before the entry of America into World War II. It's a startling document, at once naive in its conception of the wisdom that America must exert significant influence over the world, and yet in other ways, it has an honesty lacking in the pronouncements of many of our modern-day public figures.

The reader should judge whether Luce's formulation might create a more peaceful and just world, or something far different.

> As we look toward the future—our own future and the future of other nations, we are filled with foreboding. The future doesn't seem to hold anything for us except conflict, disruption, war.[28]

The answer seemed to be some kind of *Pax Americana*, a world which would be remade by America, with American values.

> And yet we also know that the sickness of the world is also our sickness. We, too, have miserably failed to solve the problems of our epoch. And nowhere in the world have man's failures been so little excusable as in the United States of America. Nowhere has the contrast between the reasonable hopes of our age and the actual facts of failure and frustration. And so now all our failures and mistakes hover like birds of ill omen over the White House, over the Capitol dome and over this printed page. Naturally, we have no peace.[29]

This writing pulls you in, engages your heart and soul, and makes you want to follow the narrator wherever he might go. Luce continues to describe the problem.

> In this self-deceit our political leaders of all shades of opinion are deeply implicated. Yet we cannot shove the blame off on them. If our leaders have deceived us, it is mainly because we ourselves have insisted on being deceived. Their deceitfulness has resulted from our own moral and intellectual confusion. In this confusion, our educators and churchmen and scientists are deeply implicated.[30]

Luce goes onto blame journalists and the media, but certainly not the American public. We simply need to be given accurate information. However, they can be moved onto a more positive path.

> In the field of national policy, the fundamental trouble with America has been, and is, that whereas their nation became in the 20th Century the most powerful and the most vital nation in the world, nevertheless Americans were unable to accommodate themselves spiritually and practically to that fact. Hence, they have failed to play their part as a world power—a failure which has had disastrous consequences for themselves and all mankind. And the cure is this: to accept wholeheartedly our duty and our opportunity as the most powerful and vital nation in the world and in consequence to exert upon the world the full impact of our influence, for such purposes as we see fit and by such means as we see fit.[31]

In other words, the world was ours for the taking.

And into the midst of these grand ambitions, the visitors came.

⋏   ⋏   ⋏

Most UFO histories begin with one of two significant events: the June 24, 1947 sighting of flying silvery discs by businessman and pilot Kenneth Arnold in Washington state, or the Roswell, New Mexico, crash of July 8, 1947, as the World War II victors, the United States and the Soviet Union, were beginning their decades-long battle for global domination. The

Roswell crash is significant in terms of national security because it took place not far from the 509th Bomb Wing at Roswell Army Airfield. The 509th was the only active nuclear wing in the United States at the time, and its fleet had flown the missions that dropped nuclear bombs on the Japanese cities of Hiroshima and Nagasaki.

But there may have been an earlier crash.

A book published in 2021, *Trinity: The Best-Kept Secret*, by longtime UFO researchers Jacques Vallée and Paola Leopizzi Harris, presents evidence of a UFO crash two years earlier, on August 16, 1945, just ten miles north of the Trinity Site in New Mexico, where mankind had first exploded a nuclear bomb, on July 16, 1945.

If true, the timeline is extraordinary.

On the morning of July 16, 1945, at the Trinity Site, Robert Oppenheimer and his team watched the first mushroom cloud explode at dawn in the desert. As Oppenheimer watched the pillar of fire rise into the sky, he recited the words of Lord Shiva from the Hindu scripture the Bhagavad Gita: "Now I am become death, the destroyer of worlds."

From that point, things moved quickly.

On August 5, 1945, the crew of the *Enola Gay* dropped the bomb nicknamed "Little Boy" on the city of Hiroshima.

On August 9, 1945, a B-29 bomber dropped the bomb nicknamed "Fat Man" on the coastal city of Nagasaki, the heart of the Japanese naval industry. It was the home of the munitions plant that manufactured the torpedoes that had sunk American ships at Pearl Harbor.

On August 15, 1945, the emperor of Japan announced his country's unconditional surrender.

World War II was over.

On August 16, 1945, an unusual object came out of the night sky and crashed on a dry, shrub-covered hill on a New Mexico ranch about ten miles away from the Trinity Site, where the first atomic bomb had recently been exploded.

According to Vallée and Harris, two young boys, Jose Padilla (nine years old at the time) and Reme (Remigio) Baca (seven years old), witnessed this crash. Harris interviewed Baca in 2010 at his home in Gig Harbor, Seattle. She also interviewed Padilla, at that time a state trooper in Rowland Heights, California, by phone. Baca explained how the two boys had instructed to ride on horseback and check the fence line for breaks, as well as for cattle that had recently calved. Prior to the crash, they had located a cow that had recently given birth to a calf.

HARRIS: (*surprised*) You heard the actual crash?

BACA: We didn't know it was a crash at the time. We heard this sound and the ground shook, and so memories came back of the atomic bomb explosion. *Are they testing again or what?* So we looked around, saw smoke coming from maybe a couple canyons down, up that way. So Jose says, "Let's go over and take a look, see what's going on." We started walking, and we saw a little smoke coming from that direction. As we reached the ridge, the smoke became intense. Then we worked our way down the ridge, *so we could see what appeared to be a big gouge in the ground.* It looked like a road grader had been in there. We weren't aware that anyone had a 100-foot-wide grader, but it sure looked like a 100-foot-wide blade, grading about a foot deep. We started walking up this road, it was pretty rough on our feet, and it was warm. *The bottom of our feet felt hot.*[32]

The two young boys were ranching kids, Padilla a descendant of Apache and early Spanish settlers, and Baca of more recent Spanish settlers. In 1945, on a ranch in New Mexico, young boys were expected to be young men, doing chores that would be considered unimaginable for their peers in modern times.

On that day in 1945, the two continued on the mysterious warm road, which made a right turn like an "L," but there was so much dust and smoke in the air from burning bushes, as well as humidity from the recent storm, that they decided to try to get to higher ground and use their binoculars.

BACA: We went back up and rested, returned, and Jose has his binoculars out and starts looking to see what it is. He says, "You know, there's something over there. Let's see if we can get any closer." Again, we try to get closer and finally it starts clearing up a little. The time seems to be going very fast. We're looking through the binoculars and I could see the hole on the side of the object. The object is avocado shaped.

HARRIS: So it's a round object like an avocado, and you could see there's a hole. How far would you say you guys were from the object?

BACA: I would estimate about a couple of hundred feet.

HARRIS: And then you saw inside of the hole from the couple of hundred feet?

BACA: No, not the inside of the hole. Jose says, "Look at this!" So I was looking through the binoculars at these little creatures moving back and forth.

HARRIS: Were they moving really fast?

BACA: They were like, sliding.

HARRIS: They were sliding…

BACA: Not sliding, but more like willing themselves from one place to another. That type of sliding. And as I'm looking at that, things begin happening in my mind.

HARRIS: Oh really…

BACA: I'm seeing them and I'm feeling this crazy stuff, like I really feel sorry for them.

HARRIS: Um, hm…

BACA: And I feel really sorry, like they're kids, too.

HARRIS: And you had a concern for them? And you're thinking…did you feel something because of the accident?

BACA: Yes, I think so. I'm hearing this high-pitched sound coming from there. We didn't know what to think. The only high-pitched sounds we were familiar with were jackrabbits when they were in pain, and also the sound of that comes out of a newborn baby when it cries.[33]

As the two boys continued to watch this scene unfold, there was a disagreement between them. Padilla wanted to investigate, but Baca was scared, saying he wanted to go home and that his parents will soon start to worry about him. They go back home, and Padilla tells his father (Faustino) about what they saw. Faustino thinks the craft probably belongs to the government, and says it's best to let them deal with it. Faustino's advice on letting the government deal with the crash was prescient, as we shall see from later events.

Although they were closest to the crash site, Baca and Padilla may not have been the first to see the downed craft. That designation may belong to Lieutenant Colonel William Brophy, as reported in *Trinity*. We have strong independent evidence that the crash did happen as the kids describe it. It comes from the crew of a B-25 on a training mission flying over Walnut Creek, who were directed to the site by Alamogordo controllers when the communication tower was destroyed. (Authors' note—One of the theories for the crash is that powerful military radars around the Roswell Army Air Base affected the guidance system of the craft.)

> The pilot, Lt. Colonel William J. Brophy, reported seeing the smoke and the bent tower. He circled the area, saw the crashed object in the vegetation and radioed back that "two little Indian boys" were close to the site.
>
> Lt. Colonel Brophy told his son in 1978 that he was the first adult witness over the site, but he didn't go near the craft on the ground until the next day, while the two kids were able to approach it shortly after the crash.[34]

In the interview, Baca told Harris the craft was approximately 25 to 30 feet long and 14 feet high, and that he knew that because when he and Padilla went back to the site, they paced it off.[35]

On Saturday, August 18, 1945, Padilla and Baca climbed into the Padilla family truck, along with Faustino Padilla and Eddie Appodaca, a state policeman and friend of the family, to examine the wreckage. They parked the truck near the crash site but couldn't see any sign of the craft, so they decided to continue on foot. They found some debris, but as they got closer, the two men told the boys to stay put, that they would investigate. Baca recalled what happened when the two men returned.

BACA: They came out and said, "Okay. Here's the way it is. I want you guys to listen. This is very difficult. You're under oath. You don't tell anybody about this, not your brother, not your cousin, not your mother, not your father, that's our business. We'll take care of that. And the reason for this is, that you can get in trouble. We want to keep you out of trouble." So we agreed to that and they gave us a really big lecture, and so we took it very serious.

HARRIS: But did they ever tell you what they saw inside?

BACA: No.[36]

Baca also recounted that it took the Army days to load the craft onto a truck and remove it from the scene.

The boys were taken home in midafternoon but were still aflame with curiosity, and they came up with a plan to get back to the site. While it had taken a long time in a truck on the roads to get close to the site, it was only about a half hour on horseback.

There were still fences to fix, as well as poles to replace, and they needed to continue to check for cows who'd given birth. They approached the site from a different direction on horseback.

BACA: Finally, we got there, we were on horseback, and came in from a different direction looking from the opposite side of the ridge, we saw some military people picking up stuff.

HARRIS: Okay. Well, that's what I had just asked you before. How did you know the military was there before? You said the creatures weren't there...

BACA: The military wasn't there all the time.

HARRIS: But the creatures were gone, and I was wondering, the military must have been there to take them?

BACA: We didn't see the military take them. If they did, it was before we arrived. But we never got to check the craft. All we got to do was go down and get some of the debris and threw it into this crevice and we tried to cover it with dirt and rocks. After the two Jeeps left, it was already getting dark and we had to get home.[37]

From Baca's account, one gets the impression that the crash caught the military by surprise, and that they were scrambling to create a plan. One or two Jeeps (Padilla's and Baca's accounts differ) with military personnel picking up material doesn't seem like an adequate response. On the following day, Sunday, August 19, 1945, a military official came to the Padilla ranch with a plan:

BACA: …And as we [he and Padillo] go in [to the Padilla home], there was a military vehicle in front and there's a soldier there at the screen door talking to his [Padilla's] dad, so we go around back and in through the kitchen to join them. Faustino says, "Come on in here, boys." So we joined him and he's talking to a Latino, Sergeant R. Avila, and he invites him in. Sergeant Avila says, "I'm with the US Army and what I need to do is get permission from you to go in and cut the fence and put in a gate, because we have one of our experimental weather balloons that inadvertently fell on your property."

HARRIS: (*laughing*) He called it a weather balloon? Those words?

BACA: An experimental weather balloon. "And so, we need to recover that, so we need your permission to do that." So his dad says, "Why can't you come in through the cattle guard like everybody else, instead of cutting my fence down?" And Avila says, "Because the equipment that we're going to bring in is wider than your cattle guard, it won't fit through there."[38]

It shouldn't come as a surprise that the military didn't have a plan for dealing with crashed vehicles filled with strange creatures. World War II had ended just a few days earlier. Was anybody expecting a potential new threat to arise so quickly? Probably not.

And it didn't seem as if anybody was concerned about the two curious boys who could get to the crash site much faster on horseback than the adults could in their trucks or Jeeps. Baca described how he and Padilla observed the military's activities over the next few days.

BACA: They were wearing fatigues, they put up a tent, played a radio, Western music.

HARRIS: You were watching them, then?

BACA: Yes, we were watching them, as often as we could, sometimes in the morning, and evening. It was our job to check and maintain fences, keep track of the herd, including horses. We could hear the radio music going. There was one guy there at the tent, and two or three working, picking up the debris.

They bring in this tractor-trailer, they have a welder, acetylene welder, and they build this rack so they can get the craft on it because it's got to go on sideways. Then we figured out they were doing that, because they had to go under the overpass at a forty-five-degree angle in order to clear it.[39]

The oddness of the scene, young boys on horseback, a crashed alien vehicle with injured creatures, and soldiers utilizing that can-do spirit American spirit paints a picture far removed from any Norman Rockwell painting. The military attempted to keep things quiet, but it doesn't seem as if they had to work very hard at it. There was trust in the US government. If the Army told you to be quiet about something, you kept quiet.

In the interview with Harris, Baca estimates that the military finished their work sometime between Thursday, August 23, and Saturday, August 25, 1945. After the military left, the boys recovered some material, including what they thought looked like a bracket (which may have been from the damaged radio tower), and kept it for several years. Later tests revealed it to be of terrestrial origin.

In the conclusion of the book, Vallée and Harris write of the difficulty of trying to verify any of the information about this incident, or others like it, over the years.

The people who hope for an imminent "Disclosure" about UFOs, and who are making valiant efforts to document the phenomenon, should take one important fact into account: *Disclosure could only come from the same organizations that are in charge of the security brief itself.* And those organizations have their own agenda and constraints, which may be regrettable but possibly legitimate.

A government entity or its classified contractors can claim to know the answers and confuse the issues with impunity. It can even come up with fake disclosures to feed the public's thirst for what passes for "truth" and the media's excitement to promote it, in the interest of what may be perceived as a higher good.[40]

If you control the secrets, it's easy to present the lies that you want. Who will call out the lies of the military, or of their defense contractors who may have been given the technology from crashed vehicles?

One might easily cite the American journalist and writer Annie Jacobsen, a 2016 Pulitzer Prize finalist, as well as a producer of the Amazon Studios streaming version of Tom Clancy's *Jack Ryan* series, starring well-known actor John Krasinski.

Jacobsen's 2011 book *Area 51: An Uncensored History of America's Top Secret Military Base* was supposed to finally give us the facts of the 1947 Roswell crash. I read it and had much the same reaction to the final secret as did Jacques Vallée. As recounted in the conclusion of *Trinity*:

> Over lunch in San Francisco, Ms. Jacobsen told me a story that was brought to her publisher by Defense Department people who stated it was all true, although it had been kept a very deep secret. The yarn went like this: "In July of 1947, Army Intelligence spearheaded the efforts to retrieve the remains of the flying disk that crashed at Roswell...and they found bodies alongside the crashed craft. These were not aliens. Nor were they consenting airmen. They were human guinea pigs."
>
> The tale goes onto "reveal" that the poor deformed children found by the Army were kept at Wright Field until 1951, and then sent to "an elite group of five [defense contractor] EG&G engineers" working at the Nevada Test Site. And what was the big truth behind the deformed aviators, child-sized with strikingly big heads and other deformities? They were kidnapped kids from Auschwitz, who fell into the hands of the evil Nazi doctor Mengele...who were subsequently re-engineered by him to turn them into grotesque child-sized aviators for Stalin,

who...then reneged on the deal to save Mengele, but kept the fake aliens to play a trick on the US and dump them on New Mexico to create a panic in America similar to what followed the [1939] radio broadcast of *The War of the Worlds*.[41]

Jacobsen goes on in her book to explain that the *real* reason we will never get the truth about the Roswell crash is that our government doesn't want to reveal that in the wake of Roswell, they also experimented on children to give them an alien-like appearance.

We have read many of Jacobsen's other books, including *Operation Paperclip: The Secret Intelligence Program that Brought Nazi Scientists to America*, about the US program to bring Nazi scientists to America; *The Pentagon's Brain*, about the Defense Advanced Research Program Agency; and *Surprise, Kill, Vanish: The Secret History of CIA Paramilitary Armies, Operators, and Assassins*, about our government's special operations teams. We find ourselves troubled by Jacobsen's books, as most of her information has the ring of truth.

But we wonder if the truths she reveals are strategic releases (accurate but no longer important) meant to lull us into believing really big lies, such as Stalin's working with Dr. Josef Mengele, the Nazi "Angel of Death," to create flying discs manned by surgically disfigured children to panic post–World War II America. (And remember, Jacobsen was a finalist for the 2016 Pulitzer Prize.)

Consider us skeptics on the Stalin-Mengele partnership as an explanation for the 1947 Roswell crash. What could be worse, or more likely to bring about world condemnation?

Logic tells us that Jacobsen is most likely a highly placed intelligence asset, or a complete fool. Which brings us to the central dilemma of this book.

If you can't trust the "secrets" revealed by a 2016 Pulitzer Prize finalist, whom can you trust?

⋏ ⋏ ⋏

What might be the truth about the Roswell crash?

Maybe it was some secret Communist flying disc manned by mutilated children.

But I'd put my money on something else.

41

In 1997, in time for the fiftieth anniversary of the Roswell crash, Pocket Books published a remarkable book, *The Day After Roswell*, by Colonel Philip J. Corso (retired) with William J. Birnes. Corso has remarkable credentials, having served as an Army intelligence officer on General Douglas MacArthur's staff in Korea, and later on President Dwight D. Eisenhower's National Security Council. After he retired from the Army in 1963, he worked on the US senate staffs of Senators James Eastland and Strom Thurmond.

In the introduction to the book, Corso reviews his military history and states that some World War II and Korean War vets were being held in the Soviet Union and Korea long after those wars had ended, that the KGB had penetrated key aspects of American foreign policy apparatus, and that rulers of the Kremlin were able to influence American policy. However, Corso shares an even bigger secret:

> But hidden beneath everything I did, at the center of a double life I led that no one knew about was a single file cabinet that I had inherited because of my intelligence background. That file held the army's deepest and most closely guarded secret: the Roswell files, the cache of debris and information an army retrieval team from the 509th Army Air Field pulled out of the wreckage of a flying disk that had crashed outside the town of Roswell in the New Mexico desert in the early morning darkness during the first week of July 1947. The Roswell file was the legacy of what happened in the hours and days after the crash when the official government coverup was put into place. As the military tried to figure out what it was that had crashed, where it had come from, and what its inhabitants' intentions were, a covert group was assembled under the leadership of the director of intelligence, Adm. Roscoe Hillenkoetter, to investigate the nature of these flying disks and collect all information about encounters with these phenomena while, at the same time, publicly and officially discounting the existence of all flying saucers. This operation has been going on, in one form or another, for fifty years amidst complete secrecy.[42]

The original manuscript had a foreword by Thurmond, which was later withdrawn. We quote from a June 5, 1997, *The New York Times* article on the controversy that erupted at the time.

> In the Foreword, Senator Thurmond, a South Carolina Republican, says Mr. Corso worked for him as an aide after leaving the Army and praises him as a person of integrity who served his country well. "He has many interesting stories to share with individuals interested in military history, espionage and the workings of our government," Senator Thurmond wrote. But he made no mention of the book's central thesis of inadvertent aid to the United States by space aliens.
>
> In a statement, Senator Thurmond said that he regretted that his foreword appeared to bolster claims of a Government conspiracy. "I know of no such 'cover-up,'" the Senator said, "and do not believe one existed."[43]

Let's apply the logic filter to this article to see if we can separate fact from fiction. Colonel Corso did work for Senator Thurmond, who was one of the longest-serving senators in United States history, serving from 1954 to shortly prior to his death in 2003, a period of approximately forty-eight years.

When Thurmond wrote the foreword for Corso in 1997, he had been serving in the Senate for about forty-three years and was one of the most powerful members of the Senate, chairing the Senate Armed Services Committee from 1995 to 1999.

Does *The New York Times* intend us to believe that Corso could work for the chairman of the Senate Armed Services Committee and be a crackpot?

Let's consider a more reasonable scenario.

Corso served on the senator's staff because he'd been thoroughly vetted as a highly skilled, trained, and experienced military intelligence officer.

Corso may not have told the senator about his experience helping reverse engineer alien craft because he didn't want to violate security oaths. But when he chose to reveal this information, the senator knew of Corso's reputation for honesty and integrity and chose to write the foreword. Are we to believe that Thurmond did not know the subject matter of the book?

Wouldn't the title itself, *The Day After Roswell*, have raised even a few questions in the Senator's mind?

We cannot know for certain why Thurmond chose to rescind his foreword. But it seems to me that there were only two possible reasons: he believed he'd made a genuine mistake, or that even as the chairman of the powerful Senate Armed Services Committee, he could not afford to offend certain powerful forces.

Let's return to Corso's book, and his description of the Roswell crash:

> Through the years, I've heard versions of the Roswell story in which campers, an archeological team, or rancher Mac Brazel found the wreckage. I've read military reports about different crashes in different locations in some proximity to the army airfield at Roswell like San Augustin and Corona and even different sites close to the town itself. All of the reports were classified, and I did not copy them or retain them for my own records after I left the army. Sometimes the dates of the crash vary from report to report, July 2 or 3 as opposed to July 4. And I've heard different people argue the dates back and forth, establishing timelines that vary from one another in details, but all agree that something crashed in the desert outside of Roswell and near enough to the army's most sensitive installations at Alamogordo and White Sands that it caused the army to react quickly and with concern as soon as they found out.[44]

Corso reports that he came into possession of the Roswell material in approximately 1960, so it makes sense that there might be some uncertainty as to the date of the crash versus the finding of the downed vehicle.

Corso then attempts to place the discovery in the larger context of the global political situation at the time.

> The military found itself fighting a two-front war, a war against the Communists who were seeking to undermine our institutions while threatening our allies and, as unbelievable as it sounds, a war against extraterrestrials, who posed an even greater threat than the Communist forces. So we used the extraterrestrials' own technology against them, feeding it out to our defense contractors and then adapting it for use in

space-related defense systems. It took us until the 1980s, but in the end we were able to deploy enough of the Strategic Defense Initiative, "Star Wars," to achieve the capability of knocking down enemy satellites, killing the electronic guidance systems of incoming enemy warheads, and disabling enemy space-craft, if we had to, to pose a threat. It was alien technology that we used, accelerated particle-beam weapons, and aircraft equipped with "Stealth" features.[45]

It must be noted that in 1997, when Corso's book was published, Corso clearly believed in the alien threat narrative. This may not have been such an unreasonable position at the time when he received these materials. But he also implicitly acknowledges that he lacked knowledge of the true intentions of these visitors.

Corso claims to have been directly involved in the technology transfer of materials from these downed alien craft to the industry leaders who attempted to reverse engineer many of the items. From his position, he was able to construct a narrative of the Roswell crash, which he provides in the book. As he tells the story:

On the evening of July 4, 1947 (though the dates may differ depending on who is telling the story), while the rest of the country was celebrating Independence Day and looking with great optimism at the costly peace that the sacrifice of its soldiers had brought, radar operators at sites around Roswell noticed that the strange objects were turning up again and looked almost as if they were changing their shapes on the screen. They were pulsating—it was the only way you could describe it—glowing more intensely and then dimly as tremendous thunderstorms broke out over the desert. Steve Arnold, posted to the Roswell airfield control tower, had never seen a blip behave like that as it darted across the screen between sweeps at speeds over a thousand miles an hour. All the while it was pulsating, throbbing almost, until, while the skies over the base exploded in a Biblical display of thunder and lightning, it arced to the lower left hand quadrant of the screen, seemed to disappear for a moment, then exploded in a brilliant white fluorescence and evaporated right before his eyes.[46]

Corso claims that this unusual intrusion set off alarm bells among the military, which suspected the object of being an enemy aircraft that had somehow crossed over from Canada or the Mexican border. The Army sent a retrieval team into the area to find the downed aircraft, while at the same time keeping civilians away from it.

But others had also seen the strange craft and understood that something had happened.

> However, finding the crash site didn't take long. A group of Indian artifact hunters camping in the scrub brush north of Roswell had also seen the pulsating light overhead, heard a burning hiss and the strange, ground shaking "thunk" of a crash nearby in the distance, and followed the sound to a group of low hills just over a rise. Before they even inspected the smoking wreckage, they radioed the crash-site location into Sheriff Wilcox's office, which dispatched the fire department to a spot about thirty-seven miles north and west of the city.[47]

The Indian artifact hunters, the Roswell fire department, and the military retrieval team from the 509th Roswell Army Air Wing arrived in short order. The Army team strung up their searchlights around the crash scene.

> In the stark light of the military searchlights, Arnold saw the entire landscape of the crash. He thought it looked more like a crash landing because the craft was intact except for a split seam running lengthwise along the side and the steep forty-five-degree angle of the craft's incline. He assumed it was a craft, even though it was like no airplane he'd ever seen. It was small, but it looked more like the flying wing shape of an old Curtis than an ellipse or saucer. And it had two tail fins on the top sides of the delta's feet that pointed up and out. He angled himself as close to the split seam of the craft as he could without stepping in front of the workers in hazardous material suits who were checking the site for radiation, and that was when he saw them in the shadow. Little dark gray figures—maybe four, four and a half feet in length, sprawled across the ground.[48]

To 2016 Pulitzer Prize–nominated author Annie Jacobsen, these were human children, surgically altered by Nazi doctor Josef Mengele, as a result of a deal with the Soviet dictator Stalin to send the United States into a state of alien panic.

However, this explanation fails to explain the incredible speed of the craft, or how it seemed to change its shape on the screen. Most of the occupants were dead, but one unfortunate was still alive. Corso recounts the chaotic scene.

> "Halt!" the sentry screamed at the small figure that had gotten up and was trying desperately to climb over the hill.
>
> "Halt!" the sentry yelled again and brought his M1 to bear. Other soldiers ran toward the hill as the figure slipped in the sand, started to slide down, caught his footing, and climbed again. The sound of soldiers locking and loading rounds in their chambers carried loud across the desert through the pre-dawn darkness.
>
> "No!" one of the officers shouted. Arnold couldn't see which one, but it was too late.
>
> There was a rolling volley of shots from the nervous soldiers, and as the small figure tried to stand, he was flung over like a rag doll and then down the hill by the rounds that tore into him. He lay motionless on the sand as the first three soldiers to reach him stood over the body, chambered new rounds, and pointed their weapons at his chest.[49]

The officer in charge was furious that the sole remaining survivor of the crash had been shot, and ordered the men to continue with the retrieval and to keep any civilians away from the site.

Although Corso claims he didn't come into possession of the Roswell materials until 1961, he asserts that a few days after the crash (he believes it to have been July 6, 1947), he saw one of the alien bodies as it was in transit through his base at Fort Riley, Kansas, on its way to Wright-Patterson Air Force Base in Ohio. The body was in a thick glass container and suspended in some kind of blue fluid.

...At first I thought it was a dead child they were shipping some-where. But this was no child. It was a four-foot human-shaped figure with arms, bizarre-looking six-fingered hands—I didn't see a thumb—thin legs and feet, and an oversized incandescent lightbulb-shaped head that looked like it was floating over a balloon gondola for a chin. I know I must have cringed at first, but then I had the urge to pull off the top of the liquid container and touch the pale grey skin. But I couldn't tell whether it was skin because it also looked like a very thin one-piece head-to-toe fabric covering the creature's flesh.

Its eyeballs must have been rolled way back in its head because I couldn't see any pupils or iris or anything that resembled a human eye. But the eye sockets themselves were oversized and almond shaped and pointed down to its tiny nose, which didn't really protrude from the skull....

The creature's skull was overgrown to the point where all of its facial features—such as they were—were arranged absolutely frontally, occupying only a small circle on the lower part of the head. The protruding ears of a human were nonexistent, its cheeks had no definition, and there were no eyebrows or any indications of facial hair.[50]

It was a serendipitous turn of events that led Corso to see that alien body shortly after the 1947 crash, fourteen years before he came into possession of the Roswell artifacts.

During those fourteen years, he served in Korea and on the National Security staff of President Eisenhower, often briefing the president on important national security issues in the Oval Office. From 1955 to 1961, Corso commanded an antiaircraft battalion in Germany that boasted the most technologically advanced surface-to-air missiles.

In 1961, with a new president, John F. Kennedy, Corso had a new assignment, courtesy of General Arthur Trudeau.

"So what's the big secret, General?" I asked my new boss. It was strange talking to a general that way, but we'd become friends while I was on Eisenhower's staff. "Why not the front door?"

[Corso had been asked to come into the General's office via a private entrance.]

"Because they're already watching you, Phil," he said, knowing exactly what kind of cold chill that would send through me. "And I'd just as soon have this conversation in private before you show up officially."

He walked me over to a set of file cabinets. "Things haven't changed that much around here since you went to Germany," he said. "We still know who our friends are and who we can trust."

I knew his code. The Cold War was at its height and there were enemies all around us: in government, within the intelligence services, and within the White House itself. Those of us in military intelligence who knew the truth about how much danger the country was in were very circumspect about what we said, even to each other, and where we said it.[51]

Trudeau and Corso believed the government had been penetrated by Soviet intelligence, and these individuals were leading the country down a destructive path. The general had a file cabinet of important information, and he knew that because of his high visibility, he couldn't be associated with the material. The general told Corso he would be working at the newly created Foreign Technology department.

The office would be just a floor below that of General Trudeau's, so that whenever the general needed him, Corso could quickly run up the back staircase and be in Trudeau's office without anybody seeing him. Corso writes:

My specific assignment was to the Research and Development Division's Foreign Technology desk, what I thought would be a pretty dry post because it mainly required me to keep up on the kinds of weapons and research our allies were doing. Read the intelligence reports, review films of weapons tests, debrief scientists and the research people at universities on what their colleagues overseas were doing, and write up proposals for weapons the army might need. It was important and had its share of cloak and dagger, but after what I'd been through in Rome chasing down Gestapo and SS Officers the Nazis left behind and the Soviet NKVD units masquerading themselves as Italian communist partisans, it seemed like a great opportunity to help General Trudeau keep some of the army's ideas out of the hands of the other military services. But then I didn't know what was inside that file cabinet.[52]

Corso's book, military service, time on Eisenhower's National Security staff, and work for Senator Strom Thurmond on the Senate Armed Services Committee leave little doubt as to his credentials.

However, Corso's next claim reveals the extent to which critical intelligence could be highly compartmentalized, even for a senior figure.

"Is there something else about this I should know, General?" I asked, trying not to show any hesitation in my voice. Business as usual, nothing out of the ordinary, nothing anybody can throw my way that I can't handle.

"Actually, Phil, the material in this cabinet is a little different from the run-of-the-mill foreign stuff we've seen up to now," he said. "I don't know if you've ever seen the intelligence on what we've got here when you were over at the White House, but before you write up any summaries maybe you should do a little research on the Roswell file."

Now I'd heard more about Roswell than I was ready to admit right on the spot my first day at the Pentagon. And there were more wild stories floating around the Pentagon about Roswell and what we were still doing there than anyone could have imagined. But I hadn't made the connection between the Roswell files and what was in the cabinet General Trudeau was talking about. Basically, I had hoped after Fort Riley that it would all go away and I could simply stick my head in the sand and worry about things I could get my brain around like bureaucratic infighting inside Washington instead of little aliens sealed inside coffins.[53]

Roswell was the bad penny that seemed to keep popping up in Corso's life. According to Corso, he would be in charge of the bulk of the Roswell materials held by the Army, which had not been exploited in the years since the crash.

> I opened the cabinet and almost immediately my heart sank. I knew, from looking at the shoebox of tangled wires and the strange cloth, from the visor-like headpiece and little wafers that looked like Ritz crackers only with broken edges and colored a dark gray, and from an assortment of other items that I couldn't even relate to the shapes and sizes of things I was familiar with, that my life was headed for a big change. Back in Kansas, that night in July, I told myself that I was seeing an illusion, something that if I wished real hard, didn't have to exist for me. Then, after I went to the White House and saw all the National Security Council members describing the "incident" and talking about the "package" and the "goods," I knew that the strange figure I'd seen floating in liquid in a casket within a casket at Fort Riley wasn't just a bad dream I could forget about. Nor could I forget about the radar anomalies at the Red Canyon missile range or the strange alerts over Ramstein air base in West Germany.[54]

The post–World War II world Corso describes depicts a community of military professionals doing their very best to avoid the full implications of what was showing up on their radar screens and being seen by their personnel.

And perhaps we should have some sympathy for these officials, as there appears to have been a genuine fear that our military had been penetrated by Communist agents, bent on thwarting the will of the United States. How many existential threats to the safety of the country could one person consider at the same time? After Corso took some time to review the materials, he received a visit from General Trudeau, checking in on his progress.

> "What did you do to me, General?" I said. "I thought we were friends."

> "That's why I gave you this, Phil," he said, but he wasn't laughing, wasn't even smiling. "You know how valuable this property

is? You know what any of the other agencies would do to get this in their hands?"

"They'd probably kill me," I said.

"They probably want to kill you anyway, but this makes them even more rabid. The air force wants it because they think it belongs to them. The navy wants it because they want anything the air force wants. The CIA wants it so they can give it to the Russians."[55]

As they continued their discussion, citing their mutual belief that the Korean War had been lost because of Communist traitors in the United States intelligence community, the plan was to keep a tight rein on this material, feeding out portions of it to military contractors so that they could develop technology that would benefit America in the event of any future conflict.

In other words, Trudeau and Corso trusted the defense contractors more than they trusted our own intelligence services.

Corso began working on an effort known as Project Horizon, which envisioned putting men on the moon to create a military platform to defend Earth against the aliens. One of the first questions Corso tackled was the nature of the aliens themselves. He eventually came to believe they'd been biologically engineered.

The medical report and supporting photographs in front of me suggested that the creature was remarkably well-adapted for space travel. For example, biological time, the Walter Reed medical examiners hypothesized, must have passed very slowly for the entity because it possessed a very slow metabolism, evidenced, they said, by the enormous capacities of the huge heart and lungs. The physiology of this thing indicated that this was not a creature whose body had to work hard to sustain it. A larger heart, my ME's report read, meant that it took fewer beats than an average human heart to drive the thin, milky, almost lymphatic-like fluid through a limited, more primitive-looking, and apparently reduced-capacity circulatory system. As a result, the biological clock beat more slowly than a human's and probably allowed the creature to travel great distances in a shorter biological time than humans.[56]

Corso goes on to describe how the hearts of the creatures were very decomposed, suggesting that our atmosphere was quite different than that of their home planet. The lungs also seemed designed to store and slowly release air in a way similar to our scuba tanks. The skeletal system also seemed to have been similarly "engineered."

> If we believed the heart and lungs seemed bioengineered for long-distance travel so, too, was the creature's skeletal tissue. Although it was in a state of advanced decomposition, the creature's bones looked to the army's medical examiners to be fibrous, actually thinner than comparable human bones such as the ribs, sternum, clavicle, and pelvis. Pathologists speculated that the bones were more flexible than human bones and had a resiliency that might be related to the function of shock absorbers. More brittle human bones might more easily shatter under the stress these alien entities must have been routinely subjected to. However, with a flexible skeletal frame, these entities appeared well-suited for potential shocks and physical traumas of extreme forces and could withstand the fractures that would cripple human space travelers in a similar environment.[57]

The picture that was developing of the alien space program was vastly different than that of humanity's early efforts with satellites and astronauts.

World War II had been fought against those who believed in the creation of a "master race" and those who did not.

These aliens appeared to have powers of bioengineering that Hitler and the Nazis would have gleefully embraced. Were these craft the scout ships of a possible alien armada, comparable to early European explorers, and this armada might be just the first wave of an invasion force infinitely more advanced than us?

Perhaps there was a good argument to be made at this time that the alien threat narrative was at least worth considering, especially if your job was defending the country. As Corso read through the reports, he found the analysis of the alien bodies even more confusing. The aliens' skin also appeared to be bioengineered; there were no food, water, or waste disposal facilities in the craft; and as best as Corso could determine, these creatures seemed to be some kind of biological robot or android.[58]

The Army had a plan for lunar exploration (actually a cover to place offensive weapons on the moon to defend against the perceived alien threat—the aforementioned Project Horizon). It would be larger than the Manhattan Project and effectively neuter the newly created National Aeronautics and Space Administration (NASA). As Corso explains:

> If the preliminary basic space station were successful, the army envisioned a more elaborate, sophisticated facility that would have its own scientific and military mission and serve as a relay station for crews on their way to or from the lunar outpost. This station would have an enhanced military capability and enable the United States to dominate the airspace over its enemies, blind its enemies' satellites, and shoot down its missiles. The army also saw the enhanced orbiting space station as another component in an elaborate defense against extraterrestrials, especially if the military were able to develop high-energy lasers and the particle-beam weapon we had seen aboard the Roswell spacecraft. The space station would, according to the army plan, effectively provide the platform for testing Earth-to-space weapons, and these, General Trudeau and I agreed, would be primarily directed against the hostile extraterrestrials who were the real threat to our planet.[59]

The Army plan for Project Horizon was Corso and General Trudeau's preferred plan. NASA provided a significant bureaucratic obstacle, but Corso and Trudeau convinced President Kennedy of the value of such an effort, even if the majority of the implementation of the plan would be handled by NASA.

> ...After his first full year in office, President Kennedy also saw the value in Project Horizon even though he was in no position to dismantle NASA or order NASA to cede control to the army for the development of a base on the moon.

> But I think we ultimately made our point to the President because he ultimately saw the value in a moon base. Shortly

after I testified before the Senate in a closed, top-secret session about how the KGB had penetrated the CIA and was actually dictating some of our intelligence estimates since before the Korean War, Attorney General Robert Kennedy, who read that secret testimony, asked me to come over to the Justice Department for a visit.

We came to a meeting of the minds that day. I know that I convinced him that the official intelligence the President was receiving through his agencies was not only faulty, it was deliberately flawed. Robert Kennedy began to see that those of us over at the Pentagon were not just a bunch of old soldiers looking for a war.[60]

Consider the minefield Corso depicts himself as having to navigate:

- He did not trust our intelligence agencies.

- President Kennedy believed his Pentagon wanted to embroil him in war with the Soviets, as the Bay of Pigs fiasco has demonstrated to him.

- Corso, as a representative of the Pentagon, was trying to show the president that the genuine threat came from the intelligence agencies, who were feeding him deliberately false information.

Attorney General Robert Kennedy read Corso's secret testimony and asked him to come to his office so they might have a discussion. Kennedy was suspicious of Corso, and vice versa. In the course of their meeting, Corso appeared to make significant headway, suggesting that the CIA was trying to mislead Kennedy, and that those like Corso at the Pentagon could be trusted.

Corso and Kennedy still did not fully trust each other. Corso did not bring up Roswell, and while Kennedy suspected there was another agenda at play, he did not push the issue. It was enough that the US mission to the moon would damage the Soviet cause, without the need for either man to bring up the more serious threat, that of the aliens having us at their mercy. Corso considered his interactions with Robert Kennedy to have been a complete success, blunting the Soviet appeal to the rest of the world while also putting us on an equal footing with the alien presence. As he says:

> The way history turned out, it was our lunar expeditions, one after the other throughout the 1960s, that not only caught the world's attention but showed all our enemies that the United States was determined to stake out its territory and defend the moon. Nobody was looking for an out-and-out war, especially the EBE's [extraterrestrial biological entities] who tried to scare us away from the moon and their own base there more times than even I know. They buzzed our ships, interfered with our communications, and sought to threaten us by their physical presence. But we continued and persevered. Ultimately, we reached the moon and sent enough manned expeditions to explore the lunar surface that they effectively challenged the EBE's for control over our own skies and sphere of space, the very sphere General Trudeau was talking about in the Project Horizon memoranda ten years earlier.[61]

In Corso's account of the secret history of the UFO phenomena, a small group of brave patriots made sure the citizens of Earth were protected against any possible alien threat. If that's true, then every person on the globe owes them a debt of thanks.

But is there another way to interpret the actions of the aliens behind the UFOs?

Might they have been observing us, and warning humanity of their concern over our tendency to violence and possession of nuclear weapons?

In Corso's account of subsequent American history, the moon landings were to demonstrate that we wouldn't be scared away from space travel, but low-level skirmishes between the aliens and the military continued. Corso claims that the US Air Force shot down an alien craft in 1974 over Ramstein Air Base in Germany,[62] and that in response:

> In 1975 and early 1976, air force nuclear weapons reposito-ries at Loring AFB in Maine, the all-important and sensitive Strategic Air Command facility at Minot, North Dakota, and other facilities in Montana, Michigan, and even the Royal Canadian Air Force Base at Falconbridge in Ontario had been

seriously encroached upon by UFOs. These weren't just random sightings. UFOs actually conducted surveillance and scanning operations at the bases that resulted in security reports and classified reports to Washington about the intrusions.[63]

The working group that was secretly dealing with the alien issue became concerned as to the aliens' true intentions regarding humanity.

These creatures weren't benevolent alien beings who had come to enlighten human beings. They were genetically altered humanoid automatons, cloned biological entities, actually, who were harvesting biological specimens on Earth for their own experimentation. As long as we were incapable of defending ourselves, we had to allow them to intrude as they wished. And that was part of what the working group had to deal with. We had negotiated a kind of surrender with them as long as we couldn't fight them. They dictated the terms because they knew what we feared the most was disclosure. Hide the truth and the truth becomes your enemy. Disclose the truth and it becomes your weapon.[64]

However, in the 1980s, with the election of Ronald Reagan to the presidency, the race to defend humanity against a possible alien threat would take a giant step forward with his "Star Wars" plan, which publicly was sold as a way to defend against a Soviet nuclear missile strike or the plans of some future madman. But in truth, it was a way to defend against the aliens, specifically through the development of high-energy lasers (HELs), which would approximate the lightning storm that had brought down the craft over Roswell.

Once launched and tested, our space-based high energy lasers, or HELs, acted like the lightning bolts on the nights of July 3 and 4, 1947, that so thoroughly disrupted the electromagnetic wave propagators in the spacecraft flying over Roswell that the pilots couldn't retain control of their own vehicle. We eventually realized that what happened then was that a natural version of an advanced particle-beam weapon burst actually brought a UFO down even as it tried to escape. When we deployed our advanced particle-beam weapon and tested it in orbit for all the see, the EBEs knew and we knew that we had a defense of the planet in place.

Gorbachev, believe it or not, was also pleased because President Reagan guaranteed that the United States would throw its defensive shield around the Soviet Union, too. Sure, the two leaders shook hands and embraced one another in public. What they had achieved together, cooperating when they were supposed to be fighting, was nothing short of miraculous. Whatever we were fighting over became minimally important in the face of a threat from creatures who were so superior to us in technology that we were their farm animals to be harvested as they pleased.[65]

In Corso's telling of the story, the Roswell crash revealed an unprecedented threat to humanity. And although the extent of that threat wasn't clear, the Army had taken the correct actions, and in what can only be described as the most fortunate of outcomes, discovered how to protect humanity against this threat.

But perhaps we should investigate another possibility—a possibility that Corso came to consider himself in the years after the publication of *The Day After Roswell*.

CHAPTER TWO

A POWER GREATER THAN NUKES OR ALIENS?

There are certain questions that are binary in nature, such as whether somebody is pregnant. One might be whether the United States government is in possession of crashed alien craft and bodies or is not.

On November 11, 2023, Michael Mazzola arranged for us to interview Paola Harris, not only about her book about the Trinity UFO crash, but about her close relationship with Colonel Philip Corso, whose book we discussed in the previous chapter.

One of the things Harris wanted me to understand is that while Corso's book had been popular among the public, Corso himself was not widely embraced by the UFO community. For many, it was likely due to the suspicion that since Corso had been on the "other side" he had been keeping the government's secrets about UFOs from the public and was likely still continuing in that capacity as part of some great deception.

In her book *Conversations With Colonel* **Corso:** *A Personal Memoir and Photo Album*, Harris explains how for the fiftieth anniversary of the Roswell crash, she was assigned to cover a large press event in Roswell, New Mexico for an Italian UFO magazine. After a long flight from her home city of Rome to Denver, Colorado, then a six-hundred-mile drive to Roswell, New Mexico, in a van with some other UFO attendees, she needed to find a hotel room. As she recounted:

> They waited for me in the van as I ran to the press area and told the people that I was a foreign journalist covering this historic event, and I needed a place to stay. They just wished me good luck and handed me the infamous yellow pages. So I placed my index finger at random on the first hotel. *The Sallyport Inn.* Sure enough, they had a room for three nights. It turned out that all the major speakers lodged there, but what was incredible

was that my room was right next to Corso's. I had no idea who he was, what he looked like, or even what he had written, but some of the best interviews have happened that way, as I am open to anything, and do not have a preconceived direction.[66]

The next day, Harris happened to hitch a ride to the morning press conference with nobody other than Corso's coauthor, William J. Birnes. Prior to Corso's talk, a young man happened to notice Paolo's Italian press badge and introduced himself as Corso's son. The son suggested that Harris should ask his father a question in Italian, as Corso was Sicilian and, after World War II, had served on the Allied Control Council, which successfully transitioned Italy from a dictatorship to a democracy without bloodshed. A 1976 article from the *Naples Daily News* recounts how Corso's Italian work had resulted in his receiving a Knight of the Crown medal from then crown prince Umberto II, who would be the last king of Italy.[67]

Corso was delighted to be asked a question in Italian (it was about the recent "alien autopsy" video purportedly about one of the Roswell crash victims). Corso responded that one particular detail had struck him as authentic and unlikely to be faked. During the autopsy, the coroner had lifted an unusual covering from the eye of the creature. While Corso never claimed to have been present at the autopsy, he recalled this detail from his reports, and said this finding had been behind the development of night vision goggles used in Vietnam. The development of this technology was one of Corso's proudest accomplishments, as he believed the device saved the lives of many soldiers in that war.

Corso later gave Harris an interview and shared stories that he did not include in *The Day After Roswell*. The first concerned what had caused the alien spaceship to crash. According to Corso, that area has some electromagnetic abnormalities, what he called "electrically charged electromagnetic pillars," which, if touched by a craft or a person, could cause them to disappear. Below are excerpts from that interview, as published in *Conversations With Colonel Corso*.

Paola Harris: Let's return to the Roswell crash. What happened with the dimensional gates in that area of New Mexico? (He had described dimensional, electromagnetic pillars when I first met him at the UFO museum.)

Corso: The night when this crash happened, [local civilian rancher] Mac Brazel said there was a lightning hit. Wilbur Smith, the Canadian physicist told me about electrically charged electromagnetic pillars. If a plane or a ship is in water, and there is a lightning hit, the nuclear binding comes apart. When one of these electromagnetic pillars is moving and something hits it, it can cause wildfires, they can become invisible. The electromagnetic binding dissolves and a human can disappear. So on that particular night in 1947, when all these storms happened and, there were three craft, and they must have come through the gate and they must have hit them, one just a fraction of a second behind another, and possibly one ship become two and appeared ten years later.[68]

Harris says that Corso went on to describe his own alien encounter with what may or may not have been one of the original Roswell craft. It happened in 1957, when he was in charge of the Army's Red Canyon missile base (near the original Trinity explosion site). A strange radar anomaly had been detected, as if something was winking in and out of existence, and Corso took a Jeep out on his own to investigate.

When he arrived at the site, he saw a UFO on the ground, appearing and disappearing. The temperature was 110 degrees, but when he put his hand on the craft, it was cool to the touch. As Harris recounts in her book:

Paola Harris: Colonel, you act like this is normal. It doesn't scare you at all?

Corso: I was in combat. If you don't act normal, you're in trouble. You go out of your mind. On the left, a being came out of an abandoned gold mine. I first pulled out the gun. Do you know what I asked the being? I asked him, "Friend or foe?" Guess what the answer was?

Paola Harris: Neither.

Corso: You are right. Neither. Wilber Smith said, "You have experienced one of the truly great events that ever happened." In a gold mine, I met one of those things. I pointed a gun at him and he wanted me to shut down my radars so he could leave. And then I put it down and asked, "What do you have to offer me?" You know what message he gave me? I'll write it down for you. He said, "A new world if you can take it."

Paola Harris: Was it like a grey?

Corso: Yes. He asked me to come aboard. I said to him, "I know what you can do to my people. Then, he asked me to shut down my radars for ten minutes. I said to myself, "If I shut down my radar, ten minutes could be an eternity." How did that thing know that I was the only man that could give that order?" I asked him, "What do you have to offer?"

Paola Harris: "A new world if you can take it," you said.

Corso: Well, I walked back from the cave, put my gun and knife away, and I walked over to the jeep. I picked up my radio and I called range headquarters and gave the order. "Capt. Williams, this is the Colonel. Shut the radar off for ten minutes. I'm on my way." But as I looked back, the thing was still at the entrance to the gold mine. What the Hell? It was standing at the entrance and there is water in there. I looked back and I saluted him.

Paola Harris: And he let you go? Sometimes they try to control you.

Corso: Yes. When I got back the sergeant said, "Colonel, you'd better pick up the range headquarters tape. Something is going 3,000 to 4,000 miles an hour on the screen." When I was there on site, I saw a green light flashing and I heard in my mind, "I'll return your salute." Was it military or was I just imagining that this thing was a soldier, too?[69]

In a postscript to the interview, Harris claims that on later occasions, Corso told her there were regular orders from above to turn off their radars for a period of time, raising suspicions that at least some high-ranking members of the military were aware that our radar often seemed to interfere with the navigation system of these craft.

In a later interview, recounted in the book, Harris asks about the possibility that these creatures were hostile, and if so, what we should do about it.

Paola Harris: Will they attack us?

Corso: There's no way in the world that I can tell you that they won't attack us. But from a soldier's point of view, I've got to be ready for any attack that might come. If I'm not, I'm negligent and I shouldn't be dressed in a military uniform. You have got to be ready for any eventuality or you'll be destroyed and your country will go with it. If they fight a war we need to realize that their super-intellect is on a different level and I don't think they would fight on our level, with bombs and cannons…. I am telling you what the British taught me in intelligence, years ago. Yes, you look for targets and things like that. But the main things you look for are "intentions." They were right. You defeat the enemy by knowing his intentions, not one bomb. It took my guesswork to build up the mosaic because I have to think in my own brain, "What is his intention?" Naturally I'm going to make mistakes because I don't know too much about him, but at least I can try.[70]

The sense one gets, from both his own book and the recollections of Paola Harris, is that Colonel Corso was an intelligent, authentic, and patriotic military man.

This is meant as a compliment of the highest regard on our part.

And yet, the passage of time can change our understanding of the wisdom of the system to which we had once pledged our loyalty. As an older man, Corso seemed to reappraise what he had done in his youth. Harris writes:

At the very end of Colonel Corso's life [Corso passed away in 1998], I remember sitting in a car with him as he was saying, "All those wars that I was in. All they did was make a predator out of me. I was a soldier; I was kind of like a hired killer." And he felt sorry that he didn't see the whole picture. He was a decorated military man, especially under General Douglas MacArthur. He was very angry about how the Korean War ended because initially it was supposed to be ideological and then it ended after a great deal of damage had been done— blowing up villages and so forth. His life was dedicated to enhancing the competitive edge of the Army, to following the orders he was given.[71]

Will any of us get to the end of our lives and believe we made all the right choices? It's a high bar to reach. Perhaps the best we can do is be honest about the choices that we've made and tell the truth as best we know it.

And yet, even if we are old, and we examine our lives as deeply as possible, there may be events for which we have no explanation. Harris had put that question to Colonel Corso in an early interview.

Paola Harris: I heard some of those fantastic war stories, but the question I have is, "How come it is you? Why you?"

Corso: I used to say, "Why me? How come? What am I doing here?" It's uncanny. My whole life is uncanny. In Rome, I didn't lose a man. None of my CIC (Counter-Intelligence Corps) agents got killed. That's impossible. It shouldn't have happened. I used to look at the orders I'd write about communist arms. That's impossible. It is not right.

I walked at times on Via Sicilia, two, three in the morning and down the street like I was in another world. Nobody around. No noise. Nothing around. Chief of police, Luigi Ferrari told me, "You have been here before. Yes. You've been here before."

So I told Gen. Trudeau "How'd I get these big positions, White House, big commands? I was drafted in the Army. I'm not a career soldier." So the general said "it doesn't matter how you got the job. We won't try to figure that out. What's important is what you do after you get the job." All my life, even now, I've always felt like somebody's been manipulating me. Who is it? I don't know.[72]

For much of his life, Corso seemed to believe that a benign force was guiding his destiny. However, as he became older, it seemed as if he questioned that belief, perhaps worrying that he had not done all he could.

Perhaps it is beyond our ability to know the answer to this question, and we must simply continue on, knowing we will make mistakes but persevering in our quest for the truth.

▲   ▲   ▲

In our interview with Paola Harris, she began by detailing her frustration that much of the American UFO community discounted the Corso story. She blamed American researchers for wanting to cash in on the Roswell stories with their books, for feeling that there was only a limited appetite for UFO stories and not wanting Corso to hog that attention.

Another frustration Harris expressed was that many of the skeptics did not appear to have done their homework to confirm Corso's military and intelligence credentials, and simply seemed to have engaged in ad hominem attacks.

One of the first significant stories she wanted to share about Corso revolved around the question she initially asked him about the Santelli alien autopsy video that had been circulating in 1997, at the time of the 50th anniversary of the Roswell crash.

> When I asked him in Italian, he told me he thought it was very real, because they had the lens. It was not the eye. The gray clone had a lens that they had peeled off. They made night vision devices from them. And Corso told me from the stage that they used to walk up and down the halls of the Pentagon [with these devices] and could see the furniture at night.
>
> Now they made these night vision devices, but remember it's the Army making them. Probably the Air Force already started with other stuff they had. But we're talking about the Army now, and they did not talk to each other. And because it's the military-industrial complex, they're going to farm their stuff out to different people, and it's about money. It's about creating a technology of night vision. The Army is not going to share it with the Air Force.
>
> From the stage, he said it was a lens. He said, "We peeled that off the eye." And actually, if you see the alien autopsy footage, you see that there is a pupil there under this closed eye.[73]

The specificity of Harris' recollection after the passage of more than twenty-five years was truly remarkable. However, she was a journalist and did publish about it, so one assumes that if her memory had become hazy about any important parts, she could have easily consulted her notes or previously published articles.

In our interview, Harris also pointed out that many people didn't understand that Corso didn't get his hands on the Roswell materials until 1960, thirteen years after the crash, and that other groups in the government may also have been utilizing the materials during that time.

Harris also believed that the Trinity crash of 1945 that she explored in her book set the stage for the well-orchestrated recovery and cover-up of the Roswell crash two years later, in 1947. It was her understanding that the Trinity crash vehicle and its occupants went to the Los Alamos lab that had exploded the atomic bomb approximately a month earlier.

She also pointed me to a controversial document, the draft of a letter to President Harry S. Truman supposedly written by atomic bomb developer and head of the Los Alamos lab Robert Oppenheimer and famed physicist Albert Einstein, in June of 1947, regarding the UFO phenomenon. The document read:

> At any rate, international law should make place for a new law on a different basis, and it might be called "Law Among Planetary Peoples," following the guidelines found in the Pentateuch [the first five books of the Bible]. Obviously, the idea of revolutionizing international law to the point where it would be capable of coping with new situations would compel us to make a change in its structure, a change so basic that it would no longer be international law, but something altogether different, so that it could no longer bear the same name.
>
> If these intelligent beings were in possession of a more or less culture, and a more or less perfect political organization, they would have an absolute right to be recognized as independent and sovereign peoples, we would have to come to an agreement with them to establish the legal regulations upon which future relationships should be based, and it would be necessary to accept many of their principles.
>
> Finally, if they should reject all peaceful cooperation and become an imminent threat to the earth, we would have the

right to legitimate defense, but only insofar as would be necessary to annul the danger.[74]

If this document was in fact written by Oppenheimer and Einstein in
1947, it's a pretty balanced summary of the potential problems. If it's a fake,
we should give it no attention.

Again, we emphatically state we cannot determine the authenticity of
this document. The most we can say is that the letter appears to be what
somebody like Oppenheimer might write if there were genuine fears about
extraterrestrial beings.

Beyond that, we're in the dark as much as any reader.

If we're to make a unique contribution to the discussion of this subject, it will be to provide a clear UFO narrative to the reader, who can then
determine if what others claim is true or false. We cannot solve a problem
until we have clearly defined it.

However, there's another wrinkle to the story of Einstein's possible
involvement with UFOs, albeit from a few years later, the summer of 1952,
when UFOs appeared in the skies over Washington, DC.

This is how a 2018 article in *The New York Times* recounted the events:

> The Washington sightings centered on events that started
> around 11:40 pm on July 19, as air traffic controllers at
> Washington National Airport noticed blips speeding near
> Andrews Air Force Base, according to government accounts.
> The unidentified aircrafts fanned out, flying over the White
> House and the U.S. Capitol. Captain [S.C. "Casey"] Pierman
> saw them that night. They vanished around 5 a.m.

> It was a second sighting a week later, though, that caused the
> wave of hysteria that forced the government to speak out.
> Albert Chop, then a spokesman with the Pentagon who was
> given the job of answering questions about the U.F.O.s, said he
> was awakened by a call on the evening of July 26....

> The Air Force dispatched jet fighters from New Castle, Del.,
> to intercept the flying objects. But every time one of the jets
> closed in, they disappeared. When the jets backed off, they
> reappeared.[75]

Can we put ourselves in the mindset of our military leaders in the summer of 1952, with President Truman in his final year of office and the former supreme allied commander of Allied forces in World War II, General Dwight D. Eisenhower, running to be president of the United States? We were locked in a cold war with the Soviet Union, fearful of that country's ambitions, fighting a hot war with the Communists in Korea, and now this new threat was in the skies over the very seat of our government?

Was this a prelude to an invasion? Or more like an angry parent storming into a noisy child's room at night and telling him to behave himself? *The New York Times* article continues:

> "It was frightening," Mr. Chop said. "I think everybody in the room was very apprehensive."
>
> At one point, a pilot found himself in the midst of four unidentified aircraft and asked what to do. "I didn't say anything," Mr. Chop told the interviewers. "Nobody said anything. All of a sudden these things began to move away from him and he said, 'They're gone!'" The pilot returned to his base.
>
> "Those things hung around all night long," Mr. Chop added.[76]

An unknown force appeared to be demonstrating its power to violate our airspace with impunity, and we were unable to do anything about it. The panic from these UFOs over Washington, DC, prompted one man, evangelical minister Reverend Louis A. Gardner, to write a letter to Albert Einstein asking for his opinion. Einstein quickly replied on the official letterhead of his office at the Institute for Advanced Study in Princeton, headed by Robert Oppenheimer.

> Dear Sir:
>
> Those people have seen <u>something</u>. What it is I do not know and I am not curious to know.
>
> Sincerely yours,
> Albert Einstein[77]

Einstein was known as the twentieth century's champion of curiosity. How could he not be curious about what had happened in the skies above Washington, DC, in the summer of 1952?

Was he scared of what he did know or what he suspected? He did not dismiss the reports. On the contrary, he supported the witnesses.

Is this evidence that Einstein had previous knowledge of UFOs, possibly from 1947 or even earlier? If there were crashed alien crafts, using propulsion technology that was unknown to us, Oppenheimer and Einstein would probably be some of the first people consulted for their opinion.

Let's move from the speculative nature of the claimed Oppenheimer/Einstein letter to a 2013 *New York* magazine article that lays out the narrative that some UFO researchers believe.

> Certainly the most contentious issue in the now 66-year history of UFOlogy, the MJ-12 saga begins with the 1947 alleged crash and recovery of an alien spacecraft outside Roswell, New Mexico. Soon after, President Harry Truman instructed Secretary of Defense James Forrestal to set up Operation Majestic Twelve, a blue-ribbon, top-secret panel headed by Vannevar Bush, a leading Manhattan Project figure and creator of the Memex machine, a forerunner of the modern-day computer. Researchers contend that the MJ-12 committee eventually brokered a sit-down between space aliens and President Dwight D. Eisenhower, during which an agreement was reached to enable alien studies of human biology via abductions and animal mutilations in exchange for use of extraterrestrial "black" technology that would lead to developments like the B-2 "Stealth Bomber." Later, it was suggested that John Kennedy's threat to reveal the MJ-12 alien negotiations was the prime reason for his assassination.[78]

If we can't determine whether something is true, the next best step is to determine if something is logical. What does not seem logical is that UFO whistleblowers are appearing in Congress only now to testify about the existence of secret military programs, when for decades the military has been telling us that UFO sightings are simply misidentifications of the planet Venus, stray weather balloons, or our personal favorite, "swamp gas."

If there were crashed alien spaceships and surveillance of our most secure nuclear sites, it seems logical that the government would mount a response while also seeking to keep such information top secret. If a secret group was established, then it would seek to exploit this technology while at the same time cloaking itself from public scrutiny.

From that point, a separate, parallel government structure would likely have developed and might resort to increasingly strong measures to conceal itself.

The 2013 *New York* magazine article goes on to note Colonel Philip Corso's book, *The Day After Roswell*, as being central to this narrative and the claim that the reverse engineering of this alien technology was responsible for the "post–World War II boom in communications technology like transistors, lasers, fiber optics, microchips, superconductors, and miracle materials like carbon fibers."[79]

In her interview with us discussing Corso's view of the government's response to the Trinity and Roswell crashes, Harris said it was important for us to understand that the alien materials went to three US military branches: the army, the navy, and the air force—and eventually NASA.

> It's because it's compartmentalized in three different military branches. And they don't talk to each other. So they're not going to say, "Guess what? We retrieved this in Brazil. Can we talk to you about what you got? What did you get, and where did you bring it?" They don't do that. And since they don't do that, nobody knows what anybody has. And nobody knows what they recovered and how they got it.
>
> So Colonel Corso was boasting. He was saying, "Yeah, we in the Army are responsible for all our technology. We don't think the Navy and the Air Force are angry." But they don't talk about it. But not everybody knows, because then it goes deep black. It goes into the military-industrial complex.[80]

Corso's book claims that alien materials were moved to civilian contracting companies because army intelligence was worried about the compromising of other areas of the United States military by Communist spies, but that created another big problem.

Namely, that these for-profit companies were now in charge of unprecedented technology.

Harris went on to acknowledge that Corso often relied on many former German scientists, who had been brought over here under Project Paperclip, as was the lead engineer for the Apollo program to get the United States to the moon, Wernher von Braun.[81]

Harris noted that Corso was immensely proud of his role in developing Kevlar, by handing some of the Roswell materials (specifically, the suit the aliens were wearing) to the University of Wyoming. She recalled that Corso talked about the atoms' all being "aligned, and it looked like a spiderweb,"[82] which caused the university to study spiderwebs, as that was the closest Earth analogue to the materials they could find.

The next thing Harris talked about was three unpublished books Corso had written, the first being called *The Dawn of the New Age*, which I could read on the Open Minds website. After the interview, I went online and read it. Harris referred to it as a book, but it read more like the type of report Corso might have written during his years in government service.

I found what he wrote to be deeply disturbing, not because of what he knew to be true, but because of what he suspected. It almost made the decades of government lies about the subject an understandable, if even defensible, response.

According to Corso, not only were the aliens able to rearrange matter on an atomic level to make the individual atoms align in the material for their suits, but they were able to do the same thing with living tissue, particularly that of the beings that accompanied the saucers. In his estimation, this made the aliens capable of quickly replacing worn-out organs with ease, even to the extent of not resulting in any blood loss. The beings that piloted the saucers were best viewed as a single unit with the craft, a realization that had eluded the researchers during the time when Corso was working on the project.

Corso also seemed to accept the claim that the Eisenhower administration had met with alien representatives, possibly at the urging of the MJ-12 group, and concluded a treaty with them. The treaty allowed the aliens to continue abducting humans and experimenting on them, as well as harvesting materials from Earth's animals, most notably cattle, in which they seemed to be especially interested. The abduction of humans and the testing, which included an interest in our genes, however, could not be done with the same lack of damage the aliens were able to achieve with the clones who flew their saucers. The aliens seemed to be indifferent to the damage that was often done to humans during these abduction experiences.

In return, the treaty allowed Americans to reengineer the aliens' technology from crashed ships, and the aliens were supposed to assist with our development of this technology. However, according to Corso, they never significantly helped the researchers attempting to reverse engineer their technologies, and any technology subsequently developed was solely the result of the effort of the human scientists.

In essence, Corso contends that the treaty Eisenhower signed was a bad deal for humanity, allowing the aliens to continue doing what they had long been doing on our planet, and not giving us anything significant in return.

In Corso's version of recent history, throughout the 1960s and 1970s, the intelligence agencies were concerned about the gap in capabilities between us and the aliens, despite not knowing whether the aliens meant any harm. Corso claims that President Reagan's "Star Wars" missile defense program was intended to defend not against the Soviet Union or rogue regimes like North Korea, but against this alien threat. And in his opinion, the program successfully developed weapons that could be used against the aliens if they were to become hostile.

Corso told Harris he had written another book, *The Day After Dallas*, which details how he had been providing information to Attorney General Robert Kennedy about this treaty, and that President Kennedy had been asking the CIA for information about UFOs in the days before his assassination, via a written memo.

To my surprise, I was easily able to find confirmation of this claim from a 2011 NBC News story on the supposed memo, which was reproduced in the article.[83] The memo reads:

TOP SECRET

November 12, 1963
MEMORANDUM FOR
The Director [redacted], Central Intelligence Agency
SUBJECT: Classification review of all UFO intelligence files affecting National Security

As I have discussed with you previously, I have initiated [redacted] and have instructed James Webb to develop a program with the Soviet Union in joint space and lunar exploration. It would be very helpful if you would have the high threat cases reviewed with the purpose of identification of bona fide as opposed to classified CIA and USAF sources. It is important that we make a clear distinction between the known and unknowns in the event the Soviets mistake our extended cooperation as a cover for intelligence gathering of their defense and space program.

When this data has been sorted out, I would like to arrange a program of data sharing with NASA where Unknowns are a factor. This will help NASA mission directors in their defensive responsibilities.

I would like an interim report on the data review no later than February 1, 1964.

John F. Kennedy[84]

The author of the article, Natalie Wolchover, goes on to state her opinion that the memo is likely to be false, but to NBC News' credit, an image of the original memo is included and credit for finding it is given to William Lester, who claims it was provided to him in 2006 or 2007 under a Freedom of Information Act he filed with the government while researching a book he was writing.[85]

Harris told us that in discussing *The Day After Dallas*, Corso supposedly told her, "You don't want to know. It's too dirty. You don't want to know.' So, I don't want to know. I don't want to know. But he wrote that book, which will never see the light of day. The other book [he wrote] was *The Man Who Saved Rome*, and that was about the CIC [Counter-Intelligence Corps in Italy, during and after World War II]."[86]

Harris also told us that she asked Corso why he was giving her all this information as well as his notes.

And I said, "Why are giving me this? Because I don't understand it."

He said to me, "Look, we figured we had them. They [the Soviets] had them. Bodies of these clones or beings used to fly the ships." His thinking was, "These are used to fly the ship. We're afraid of the guys that made them."

And I said, "Why are you afraid of the guys that made them?"

And he said, "Because they look like us, and they could be walking in the halls of the Pentagon."[87]

Despite his analysis as a military man that the aliens could pose some threat to humanity, Corso's personal view was that they were not hostile. As Harris recounted:

> He thought they were evolved beings and that we were at the bottom of the barrel. And you asked how many races [he thought were visiting us]. He said at the Pentagon, they believed it was fifty-two different species. Some of them don't even look—I mean, when I talked to Clifford Stone [whose job was supposedly to assist with downed aliens and crash retrieval], he said it was fifty-seven. The ones in Vietnam looked like frogs. They were amphibian-looking. And Clifford had a manual with fifty-seven species—he said there was fifty-seven—because they had to sometimes apply first aid. And if some were hurt, you couldn't put alcohol on them or something. And I didn't realize with crash retrieval, you could also help them.[88]

Just when we thought we were getting used to the idea of one alien species visiting our planet, Harris blew our minds by suggesting the true number was somewhere between fifty-two and fifty-seven species.

But Harris didn't seem to be as worried about the aliens as much as about humanity, particularly the UFO community, for whom she had harsh criticism.

> They haven't put it together, because they don't read each other's books. They only read their own and promote their own. They have their own opinions. They put it in a box and don't go out of their box. And there's no logical dialogue. If you do some studying and read a whole bunch of books, we can have a dialogue, because we have material in common that we've read. I can't have a conversation with any of those guys, because they haven't read anything. They don't know the dates. They don't

know that NATO was having problems with the Berlin Airlift in Germany during the Cold War—Colonel Corso told me this. He said, "We [the government] don't have to worry about the UFO community. They fight among themselves. We don't have to put this information [fake stories] out there [to make them fight]. They do it to themselves."[89]

Her observations ring true to us. Some people like to fight with an informed point of view, and others simply like to fight. A prominent activist in the health freedom community once told Heckenlively, "I spend half my time fighting the bad guys on the other side, and half my time fighting the crazies on my own side."

In the UFO community, Harris greatly admired Steven Greer, but has little use for whistleblower David Grusch, whom she feels was only repeating things he'd been told by others. She believes that the story of the aliens who are involved with humanity is the most important story of all time, and it's not being told correctly, especially the part referred to as "abductions." As she told us:

I think that the field itself has become entertainment, where you go to a conference and say you've been abducted, and you become important. And you've got so much psychological baggage behind you that it makes you special. These hypnotherapists have come up to me and said they can't pay their bills unless they convince people they've been abducted. [We took that to mean patients will only pay therapists if the patients believe the experience was against their will.]

And I say, "No, change the word to 'contacted.'" Because I think a lot of people have been contacted and not abducted.

The thing is, I see the field as different. I see it as visitations or contact with planet Earth in an attempt to evolve consciousness and the species itself. But our human nature makes it entertainment: our own shows, separation, [and] wanting to keep material, whether it be three branches of government or the infighting in the UFO community. None of this is done properly. It's not done on a scholarly level, the way you would study geology, psychology, or any of the humanities. You wouldn't be reading your own book to come to a conclusion. I have a library of three hundred books, and I've read them all.

I'm going to figure it out. It's a puzzle that needs to be put together.[90]

Perhaps we should remember the words Colonel Corso told Paola Harris near the end of his life, about how he felt he was being "guided" by something. According to him, this force protected him in war, as well as during some of the most consequential events of the twentieth century.

Hopefully, there was a "force" guiding Corso, even protecting him.

And in parallel, we must hope that this same force guides and protects us, as we make our way through this maze of smoke and mirrors.

CHAPTER THREE

THE MAURY ISLAND INCIDENT, A MAN IN BLACK, A PLANE CRASH, AND THE KENNEDY ASSASSINATION

As noted earlier, the birth of the modern UFO age is usually dated June 24, 1947, when private pilot Kenneth Arnold claimed to have seen nine shiny discs near Mount Rainier, Washington, flying in formation. He described each craft as being about a hundred feet across, with no discernable tail, and observed that they weaved from side to side, banked, and sometimes flipped. He estimated their speed to have been approximately twelve hundred miles per hour, and this was several months before pilot Chuck Yeager broke the sound barrier by going seven hundred miles per hour.

However, Arnold wrote about an earlier incident, which has come to be known as the Maury Island incident, in his 1952 book *The Coming of the Saucers*. It took place three days before Arnold's sighting, in the same general area of Washington state. In addition to being one of the first of the "modern" UFO sightings, this might mark the first appearance of mysterious figures, either from the military or what has become known as the "men in black," and their potential origin.

UFO researcher Richard Dolan summarized the story in his book *UFOs and the National Security State: Chronology of a Coverup 1941–1973*:

> The incident was said to have occurred on June 21, 1947. Harold A. Dahl, a log salvager on Maury Island (situated in Puget Sound between Tacoma and Seattle) who helped the local Harbor Patrol Association, who was out on the bay with his son, dog, and two crewmen. He saw six large, metallic, doughnut-shaped aircraft, about one hundred feet in diameter and two thousand feet above. Five objects circled around one that seemed to be in trouble and losing altitude. Dahl heard no

sound, and saw no motors, propellors, or means of propulsion. The objects had large, round portholes on the outside and a dark, continuous, "observation" window toward the bottom and inside. As Dahl took three or four photographs, one craft moved toward the center, apparently to help the troubled craft. A dull explosion followed, and the troubled craft ejected a stream of light metal which "seemed like thousands of newspapers," then ejected a heavier and darker, similar to lava rock. After this, the craft lifted slowly and drifted out over the Pacific Ocean, disappearing from sight. Dahl said the heavy material damaged his boat, killed his dog, and injured the arm of his son, requiring a trip to the hospital. He described the event to a man he described as his supervisor, Fred L. Crisman, estimating that twenty tons of material had fallen.

Actually, Fred Crisman was an intelligence agent, formerly of the OSS and soon to be CIA, who specialized in internal "disruption" activities. This fact was unknown for many years, then suspected, then finally proven with the discovery of certain CIA documents.[91]

When one puts this story together with the Trinity crash two years earlier and the Roswell crash that would take place a few weeks later, one is struck by the sense that whoever was engaging in these activities was doing it in something of an emergency situation, possibly without having everything properly prepared.

The fact that Dahl reported this event to Fred Crisman, a man who would later be revealed to be an intelligence agent, is highly suspicious, as is what followed.

Dahl also claimed that the next morning an ordinary-looking man in a black suit arrived at his house and invited him to breakfast. This was not as unusual as it might seem. Many lumber buyers called on people in Dahl's business to buy salvaged logs. Dahl followed him to a diner, ordered breakfast, and listened in astonishment as the man related Dahl's entire experience from the day before, all in precise detail. "What I have said to you is proof to you," said the man, "that I know a great deal more about this experience of yours than you will want to believe." He warned Dahl not to discuss the experience. Dahl considered the man to be a crackpot and mailed some fragments to Raymond Palmer, a Chicago publisher of the paranormal.

Around July 22, 1947, Palmer wrote to Kenneth Arnold [about four weeks after Arnold's sighting and about two and a half weeks after the Roswell crash], and asked Arnold whether Arnold would look into it and send back some fragments.[92]

For the modern reader, understanding the amount of suspicious activity going on in the summer of 1947 is vital, from Harold Dahl's encounter on June 21, 1947, to Kenneth Arnold's sighting on June 24, 1947, to the Roswell crash on July 3 or 4, 1947.

A 2014 book titled *The Maury Island UFO Incident*, by Charlette LeFevre and Philip Lipson, gives some added description of Dahl's mysterious morning visitor.

The man, "wore a black suit, was of medium height." He had been driving a new "1947 Buick sedan." The man invited him to a nook breakfast café in the "uptown section" of Tacoma and began to relate in "great detail" the experience that Harold and his crew had seen, as if he had been there himself. The man made strong, not-so-veiled threats and told Dahl he and his crew "had made an observation that shouldn't have happened" and that "if he loved his family and didn't want anything to happen to his general welfare, he would not discuss his experience." Dahl would later relate he didn't put much stock in what the fellow said and didn't intend to keep his experience a secret, later discussing the event with other seamen at the pier.[93]

In this account, it's difficult not to discern the beginnings of a familiar, and yet disturbing, narrative.

Who was this man in a black suit who suddenly appeared at Dahl's doorway?

In *UFOs and the National Security State*, Dolan goes on to note some other disturbing aspects of the Maury Island case, such as Arnold's later being visited by two representatives of military intelligence who were in the Fourth

Air Force, Lieutenant Frank M. Brown and Captain William Davidson, as well as Ray Palmer of *Fate* magazine, wiring Arnold two hundred dollars through Western Union as payment for investigating the case.[94] (Arnold and Palmer would later collaborate together on the book *The Coming of the Saucers*.) As Arnold investigated the case, the strangeness intensified, according to Dolan:

> Arnold arrived at the hotel and found Harold A. Dahl in the phone book. After some determined prodding by Arnold, Dahl visited Arnold that night, then took Arnold to his "secretary's" home to see some fragments, one of which served as an ashtray. Arnold said it looked like simple lava rock. No, said Dahl, this was the stuff that hit his boat, and Crisman had a box of it in his garage. The next morning, Arnold met with Dahl and Crisman, who claimed to have had an independent flying saucer sighting. Arnold asked Dahl for the photographs and Crisman for the fragments. Feeling that something was wrong but distrusting his ability to evaluate the situation, Arnold then called his friend, pilot E. J. Smith ("Smithy") who had also witnessed a flying saucer. Smith arrived that day, cross-examined Dahl and Crisman, but could not trip them up. The two investigators then decided that Smith would stay in Arnold's room, and that they would see everything the next morning, including Maury Island.[95]

After interviewing Dahl and Crisman, Arnold received a call from a United Press reporter, Ted Morello, who told Arnold that a source had called him and told him everything that had happened during Arnold's questioning of Dahl and Crisman. Although Arnold suspected the leak was Dahl or Crisman, the reporter eventually informed Arnold that it was his supposed friend, the pilot E. J. Smith ("Smithy").[96]

Why would Arnold's friend do such a thing?

The circle of individuals involved in this incident was still quite small, but about half of them appeared to have questionable motives.

There was Harold Dahl and Kenneth Arnold, who should be characterized as the first two significant UFO witnesses of the modern era.

Then there was Arnold's friend, E. J. Smith ("Smithy"), who didn't seem to be much of a friend, since he called United Press reporter Tom Morello and told him exactly what their investigation was revealing.

Next is Dahl's boss, Fred Crisman, who it turns out was an intelligence agent (and who would in later years be accused of being involved in some very unsavory activities).

And finally, there were two representatives of military intelligence from the Fourth Air Force, Lieutenant Frank M. Brown and Captain William Davidson.

Dolan's account of the Maury Island incident continues:

> On the morning of Thursday, July 31, Crisman and Dahl brought heavy fragments and some white metal. The lava-like pieces were unusually heavy, smooth on one side, and slightly curved. On the other side they looked as though they had been subjected to extreme heat. The white metal had square rivets as opposed to the standard round ones, but seemed normal otherwise. Regarding the photographs, Dahl said he had given the camera with its film to Crisman, who could not now find it, but would try in the afternoon.[97]

Arnold was then in contact with Brown and learned that the plan was for the material to be loaded onto a B-25 for transport to a military base for inspection. Arnold was insistent that Brown call from an off-base pay phone rather than from on the base.[98] Dolan picks up the account from July 31, 1947:

> Brown and Davidson arrived in the late afternoon, and the five men (Dahl had left) talked until 11 p.m., at which time Crisman offered to go home and retrieve another box of fragments. No thanks, said the officers, they were no longer interested. Anyway, they had to return for Air Force Day, the next morning. Every plane on the base, including their newly overhauled B-25, had to be ready for maneuvers....
>
> At 1:30 A.M., Brown and Davidson's plane exploded and crashed, some twenty minutes after taking off. Also aboard were an army hitchhiker and engineer who survived. According to their account, the two officers had loaded a heavy cardboard

box. Shortly after takeoff, the left engine caught fire, the emergency fire-fighting system failed, and Brown ordered the two men to parachute. For the next ten minutes, they watched the burning plane continue to fly. The army blamed the crash on the loss of an exhaust collector ring on the left engine but could not answer why Brown and Davidson failed to signal distress or bail out.[99]

This sounds like a remarkable string of coincidences: that a plane carrying what appeared to be definitive proof of something unknown traversing our skies caught fire, and the men in charge of the materials had the presence of mind to tell two of their passengers to parachute to safety but failed to do the same.

On August 2, 1947, an article appeared in the *Tacoma Times*, written by investigative journalist Paul Lantz, who himself would be dead within six months, at the age of twenty-nine. It included the following:

> The mystery of the "Flying Saucers" soared into prominence again when the Tacoma Times was informed that the crash of an army plane at Kelso may have been caused by sabotage.
>
> The Times informant, in a series of mysterious phone calls reported that the ship [aircraft?] had been sabotaged or 'shot down' to prevent shipment of flying disc fragments to Hamilton Field, California for analysis.
>
> The disk parts were said by the informant to be those from one of the mysterious platters, which plunged to earth on Maury Island recently.[100]

Of some historical interest might be that the United States Air Force officially became its own branch of the military on August 1, 1947, which means that the crash of the B-25 in the early morning hours of August 1, allegedly carrying extraterrestrial materials meant for Hamilton Air Force Base in California, was the first fatality of the newly formed service. The CIA would be formed a few weeks later, on September 18, 1947.

If one takes as fact the incident at Maury Island, the Kenneth Arnold sighting, and the Roswell crash, all in the summer of 1947, it isn't difficult to understand why the agents in these services would be concerned about UFOs from the very beginning of their existence.

It would be a dereliction of their duty to this country not to place these issues at the very highest level of concern.

While the two military investigators, Davidson and Brown, had died in the crash, journalist Paul Lanz, twenty-nine years old at the time, was pursuing the mystery.

However, he would die on January 10, 1948, supposedly of Streptococcal meningitis, which didn't seem like a good explanation to his surviving family.[101]

For their book on the Maury Island incident, LeFevre and Lipson interviewed Lanz's granddaughter about her grandmother's (Lanz's wife's) memory of two FBI agents visiting Lanz about his reporting on the flying saucer story. The granddaughter says:

> The only details that I have about this visit (so far)...are that there were two of them dressed in black, she recalls... They came in showing their FBI identification.... My grandma went into the kitchen to cook and clean while they talked to Paul in the living room. She tried to listen in on their conversation.... She said they were basically telling Paul to stop. She says they were threatening him...but Paul was bold and clearly not afraid of them in any way....
>
> She said they were there for what seemed like hours, but in reality, maybe only a couple of hours. She said that Paul was defiant and determined to get to the bottom of things and he was not going to let the FBI stop him from doing so![102]

This early account is genuinely puzzling—the intimidation of somebody investigating a UFO report and the crash of a military plane, with FBI agents visiting him and warning him to stay away from the case.

One gets the sense that something enormously important is being concealed, and yet it eludes us.

The interest of the FBI would make sense if these vehicles were some weapon of the Soviet Union, designed to panic us. But since nothing remotely like that has been reported in the ensuing decades, it seems unlikely. It also seems strange that there would be such a quick federal response, but we'll simply leave it at that. However, the FBI was not done with this case. As Dolan wrote:

> On August 7, 1947, the Seattle FBI office interrogated Crisman and Dahl, and soon announced the affair had been a hoax, and the two had also been the mysterious informants. They supposedly offered Palmer their exclusive story for money—something Palmer vehemently denied. The fragments mailed by Dahl, said the FBI, were not from a flying saucer. Palmer, by the way, claimed that the fragments were stolen from his office shortly thereafter. Before they were stolen, he had them sent out for analysis. The results, he said, indicated that they were neither slag nor natural rock.[103]

There is a binary choice in this story.

On one side is believing these people and their accounts, maybe even marveling at some of the consistencies in several other stories like it. The other choice is to believe that living among us are people simply waiting for the opportunity to create a hoax that will permanently change the way family, friends, and others they have yet to meet will forever treat them.

We may not have the skill to peer into the souls of others, but we do have a pretty good sense of which possibility is more likely.

However, as much as we may be confused about this case, it also seems to have caused confusion at the top level of the FBI, as described in Dolan's book.

> Behind the scenes, the matter was not so clearly settled. J. Edgar Hoover wrote the following week: "It would also appear that Dahl and Crisman did not admit the hoax to the army officers…" In response, the FBI special agent in charge from Seattle answered:

> > Please be advised that Dahl did not admit to Brown that his story was a hoax but only stated that if questioned by authorities he was going to say it was a hoax because he did not want any further trouble in the matter.[104]

J. Edgar Hoover seemed to be just as confused as others were about the situation.

Consider this next piece of information. While Harold Dahl seems to have been an innocent in this affair, consider what was later learned about his superior, Fred Crisman. This is a short summary about Crisman from LeFevre and Lipson's book on the Maury Island incident.

> Crisman would become one of the most interesting characters in the Maury Island UFO incident with even more covert relationships than one would think a Tacoma small businessman and teacher would have.

> FBI documents and statements from friends would show Crisman would later be involved with political campaigns, shady church organizations, as well as eventually be subpoenaed by Jim Garrison on the JFK assassination. It is very likely much of his skills played a part in manipulation and coverup in the Maury Island UFO incident.[105]

Adding to the mystery, a year before the Maury Island incident, Crisman was publicly talking about UFOs, specifically in a letter he wrote to the magazine *Amazing Stories*, as recounted by a writer on Medium.

> Our story begins in March of 1946, when *Amazing Stories* magazine published the first of what would become known as the Shaver mysteries.

> A writer named Richard Shaver began publishing "true" stories of his former life in the lost continent of Lemuria. The most shocking part of his story was the claim that remnants of Lemuria still exist today under the ground.

> In dark caverns there are said to be savage, immoral beings called the Deros. They had access to Lemurian technology, such as ray guns.[106]

Crisman was furious that such stories were being published, and in a letter to *Amazing Stories*, alleged he'd been in a gun battle with aliens in a cave in Burma in 1945.[107]

No matter how we look at this piece of information, we can't make any logical sense of it. Let's assume for a moment that Crisman is telling the truth, and he was engaged in a gun battle with aliens in a cave in Burma in 1945.

It strains credibility to believe that he would find himself in a similar position in 1947, giving information to the FBI about one of the first significant modern UFO cases.

The much more plausible explanation is that Crisman was a disinformation expert, creating the sense that aliens were real in his letter to *Amazing Stories* in 1946, then shaping the new UFO narrative in 1947, even to the point of lying to Hoover's FBI.

We'd be more inclined to believe this narrative if I didn't have to add to it the bringing down of an Air Force plane and the death of two service members. (There is also a third possibility that the plane crashed for reasons unrelated to the actions of the intelligence agencies or aliens, and the intelligence agencies simply tried to take advantage of it.)

But we get a little concerned when too many coincidences begin to pile up.

According to records, Crisman flew 211 combat missions in World War II, then after the war went to work for the Department of Veterans Affairs, helping those servicemen who found themselves in trouble with the law.[108]

There was a report to the FBI that Harold Dahl was some kind of "black-market operative" and that it was during Crisman's time working for the Department of Veterans Affairs that he became associated with Dahl, although it is difficult to determine their exact relationship. From the book on the Maury Island incident:

> Crisman told an informant that he had developed an idea for a log patrol and beach patrol, which would involve the recovery of unmarked logs from Puget Sound and the patrolling of summer beach cottages for private owners. Crisman said that Dahl stole this idea from him.

> A few days after the Kelso crash, Crisman was ordered by the Army to Alaska. There is no evidence of Crisman actually going. A statement given to the FBI on August 8 led the Fourth Air Force headquarters to revoke his Air Force Reserve Commission as an "undesirable and unreliable officer."[109]

What are we to believe about Crisman? Given his exemplary war service, the revocation of his commission seems confusing. Could this have been a way to release him from military service to join the intelligence agencies?

Adding to the confusion, Crisman was recalled to active service during the Korean War of 1950 to 1953, in which he served as a fighter pilot for two and a half years before returning to teaching.[110] Later, he worked for two years for the Boeing aircraft company as a sales representative, as well as had numerous other businesses and jobs, such as being a radio host as well as being a writer for several local organizations.

But Crisman gained the most attention of his life in the 1967 investigation by New Orleans district attorney Jim Garrison into the JFK assassination (dramatized in the 1990 movie *JFK*, by Oliver Stone).

Regardless of the publicity Crisman would receive during the Maury Island incident; it is his history later in life and with the JFK assassination investigation by Jim Garrison where he would achieve the most notoriety during his life. Crisman would be the first person [businessman and military officer] Clay Shaw would contact after his arrest by Jim Garrison for conspiring in the assassination of Pres. John F. Kennedy. This contact strongly implies the Crisman and Shaw were CIA operatives [Shaw was later confirmed as a CIA operative, after his death] and that somehow Crisman was his supervisor or could somehow get him out of this situation.

Crisman himself would later be subpoenaed by the Orleans Grand Jury in January of 1968 and asked what his relationship was to Thomas Beckham—a suspect in the JFK assassination. [In his testimony, Crisman said he knew Beckham held money for anti-Castro groups, and after the Kennedy administration shut down their activities, Beckham may have used that money to finance the assassination.] Crisman, in his statement to the Grand Jury said he met Beckham through Harold Dahl in 1966 [the two were apparently still friends nearly twenty years after the Maury Island UFO incident]—that Harold Dahl, who was operating a small secondhand store, had introduced Thomas Beckham to Crisman. The facts are that Crisman met Beckham years earlier, in the summer of 1963.[111]

In his testimony before the New Orleans grand jury that Garrison convened, Crisman depicted himself as a businessman trying to help Beckham with a recording career.

But who was helping whom?

Beckham would later write a book titled *Remnants of Truth: Revealing Evidence on the Jim Garrison Investigation,* in which he confessed to lying to Garrison and his investigative team. As LeFevre and Lipson write:

> Beckham believed he was used in some type of plot. He said if he knew now, what he knew then, he would have run. "But as far as I ran, someone always managed to find me (like my CIA buddies Jack Martin and Fred Crisman)." Beckham relates in his book how right after his brother contacted him about the newspaper's announcement of his subpoena that "Within ten minutes or even less, Jack Martin and Fred Crisman called me about the same thing." Crisman and Martin told Beckham not to worry, "everything was being taken care of."[112]

And what did district attorney Jim Garrison believe about Fred Crisman? According to documents that Garrison filed with the House Select Committee on Assassinations (HSCA), Garrison believed Crisman to be deeply involved in the assassination, not only as the organizer but also as a possible shooter (one of the three "tramps" taken into custody by the Dallas police but not booked), and that Crisman's skill as a pilot may have enabled him to quickly escape back to Washington state. This is part of a memorandum Garrison submitted to the HSCA in 1977:

> The appearance of CLAY SHAW in San Francisco on November 22, 1963, and in Portland, Oregon on November 25, 1963, appear to be significant—and an understanding of the surrounding circumstances will be helpful in the investigation, notwithstanding that fact that SHAW is deceased.

> It will be recalled that at DAVID FERRIE's apartment—during the pre-assassination conversation participated in by SHAW, FERRIE and "LEON" OSWALD, witnessed by PERRY RUSSO, SHAW stated he was going to arrange that on the day of the assassination he would be making a speech somewhere.[113]

Crisman lived in the Portland, Oregon, area, so Shaw's presence on the West Coast, right after the assassination was suspicious. Garrison further writes in the same memo:

III. Crisman Residency Factors with Regard to Oregon

Reference is made to you in the letter to you in which I indicated that a very productive working hypothesis at this point would be to recognize the indications that FRED LEE CRISMAN seems to have functioned as some form of "cut-out" for anti-Castro operations (and the subsequent outgrowth of a part of that: the assassination operation, itself) in the Southeast sector of the country, and in New Orleans, in particular. A "cut-out" as you will appreciate, is in intelligence terminology, an insulated contact for an operation between higher level decision makers or other individuals and lower echelons where the day-to-day machinery is actually in motion. Essentially, his role is that of a key contact man. His function may be supervision or it may merely be monitory with regard to a particular mission. One of his customary chosen assets for such an assignment will be his personal geographic remoteness from the general area of operation—hence the term "cut-out." His prompt removability afterwards from the area where the operation was accomplished provides—especially when the circumstances are sensitive—a high degree for the group or clique using him as its "cut-out" and helps the group keep the enterprise clandestine and frustrating to regular inquiry procedures.

I have already indicated, in my previous correspondence and memoranda, why I believe CRISMAN deserves the honor of being a high priority investigative target. I am suggesting here that SHAW's November 25th and November 26th visit to Portland justifies close examination and makes the CRISMAN potential all the more interesting....

Admittedly, even here—if CRISMAN is, as he rather clearly appears to be, some form of domestic intelligence operative— bells may be rung by too premature, and too blunt an inquiry. However, that question of timing is one to be determined by you and your staff, and it may be possible to develop, through indirection, what the Bureau describes as a "pretense" cover for the record inquiry which would have to be made to the Pentagon (or, perhaps, directly to the military records department at St. Louis).[114]

Some have claimed that Garrison came the closest to the truth of all the law enforcement individuals about the assassination of President Kennedy. To others, Garrison was a figure who made wild accusations unsupported by fact.

But what is true without a shadow of a doubt is that Crisman was at ground zero of two of the most important events of the twentieth century: the beginning of the modern UFO era and the investigation into the assassination of President Kennedy.

Our brains are hardwired to resist fantastical tales, and yet even a conservative reading of this story reveals an unlikely chain of coincidences.

A World War II pilot claims that near the end of the war, in 1945, he engaged in a gun battle in a cave in Burma with alien beings.

In 1947, this same man is involved in probably the first significant UFO sighting of the modern era, which is accompanied by physical evidence, which vanishes when the Air Force plane carrying the material crashes after takeoff.

When questioned by the FBI, Fred Crisman and Harold Dahl say they will tell people it was all a hoax.

J. Edgar Hoover, head of the FBI, can't figure out whether the sighting was real or not.

Shortly after the incident, Crisman is relieved of his reserve status in the Air Force.

When the Korean War starts in 1950, Crisman is called back into the Air Force.

After Crisman's second round in the service, he becomes a teacher as well as spends some time working for the Boeing Aircraft company and pursuing other businesses, while keeping up his friendship with Harold Dahl.

In the summer of 1963, Dahl introduces Crisman to Thomas Beckham, who is responsible for holding money for anti-Castro groups.

When Clay Shaw is arrested in New Orleans for the murder of President Kennedy, the first person he calls is Fred Crisman.

There are a lot of facts about Crisman, many of them contradictory, and we don't feel we can bring the picture of him into clear focus. We're just as confused about Crisman's real identity and motives as were New Orleans district attorney Jim Garrison and FBI director J. Edgar Hoover.

▲ ▲ ▲

Somebody who presents a less complicated picture than Fred Crisman is pilot Kenneth Arnold.

In January of 2023, Arnold's granddaughter, Shanelle Schanz, republished his original book, *The Coming of the Saucers,* with some additional material. She wrote of her grandfather in the foreword:

> My grandfather's sighting changed the world. He was the right person, at the right time. A pilot's opinion had merit and faith. Modern Ufology was born as a new science. He had more sightings of 'flying saucers' in his life as a pilot which he kept private....
>
> My grandfather told me before he died, *"always think for myself and not to trust the government."* I was always proud of my grandfather when he told my 2nd grade teacher who he was. And it was in the 1980s when talking about UFOs was still taboo. He died disappointed that the American government never gave a definite answer into what the flying saucers were.[115]

It's a nice foreword, and it's clear that Schanz dearly loved her grandfather. But who was Kenneth Arnold? Let's hear from him in his own words, from *The Coming of the Saucers.*

> What I saw over the Cascades in the State of Washington, as impossible as it may seem, is fact. I never asked, wanted, nor expected any notoriety for accidentally being in the right spot at the right time to observe that chain of nine mysterious objects. I reported something any pilot would have reported. If, reasoning along patriotic lines alone, I had not reported my observations, I would have been rightfully considered disloyal to my country. My observations were not due to any particu-

lar sensitivity of eyesight, nor to any abnormal or supernatural ability....

Since that day, I have been repeatedly questioned and investigated by such agencies as Military Intelligence, the Federal Bureau of Investigation, Bureau of Internal Revenue, Central Intelligence, Marine Intelligence, private detective agencies, individuals, and just plain busybodies.[116]

Early in Arnold's book is an artist's rendering of the mystery craft he saw, and it can be shocking to modern eyes. It does not look like our traditional image of a flying saucer; it resembles the profile of a current B-2 bomber, although there is a central orb structure in the middle of the craft, where one might expect a pilot to sit.

The other thing that comes across in Arnold's account is an earnest patriotism that is almost shocking to encounter in our modern cynical times. He considered his reporting to be along "patriotic lines" that any good American would understand, and that a failure to make a report would have been "disloyal" to his country.

The vast majority of us do not think, let alone speak, in such terms.

He ends his introduction with much the same hope we have in writing this book, seventy-seven years after Arnold's sighting.

To me, the search has not ended. But participation of the sort involved in the fatal "Tacoma Incident" in which two Central Intelligence military investigators lost their lives and a hundred-thousand-dollar aircraft was destroyed is something that I would refuse to enter into again—largely because it is my conviction that nothing concrete will be learned by such obtuse methods. This thing is big. It is something to which we most certainly ought to have the answer. It is my conviction that the facts already at hand must not be buried beneath a mass of official stupidity and a smokescreen of variant idiocies. It is time the truth was sifted from the untruth, the facts from the fakery, the real flying saucer from the fanciful.[117]

When seeking to unravel the sightings of Harold Dahl and Kenneth Arnold, we find that there seem to be three distinct groups: the first is composed of seemingly regular citizens reporting something strange; the second, a cabal of military and intelligence people who often appear to have conflicting agendas for good or ill; and the third, something else—a distinct group.

We think it's important to examine Arnold's own words about the incident that took place on June 24, 1947, as he was flying through the Cascade mountain range, hoping to spot the location of a downed C-46 Marine transport that had disappeared with thirty-two Marines. The military had posted a five-thousand-dollar reward for the pilot who discovered the crash site, and Arnold was interested in that bounty.

> It was during this search and while making a turn of 180 degrees over Mineral, Washington, at approximately 9200 feet altitude, that a tremendously bright flash lit up the surfaces of my aircraft. I was startled. I thought I was very close to collision with some other aircraft whose approach I had not noted. I spent the next twenty to thirty seconds urgently searching the sky all around—to the sides above and below me—in an attempt to determine where the flash of light had come from. The only actual plane I saw was a DC-4 far to my left and rear, apparently on its San Francisco to Seattle run. My momentary explanation to myself was that some lieutenant in a P-51 had given me a buzz job across my nose and that it was the sun reflecting from the surface of his wings as he passed that had caused the flash.

> Before I had time to collect my thoughts or to find any close aircraft, the flash happened again. This time I caught the direction from which it had come. I observed, far to my left and north, a formation of very bright objects coming from the vicinity of Mount Baker, flying very close to the mountain tops and traveling at tremendous speeds.[118]

The picture Arnold paints is of a regular man, a skilled pilot, engaged in one task when something completely unexpected happens. While Arnold could not at first make out the objects' shapes, he estimated them as being approximately a hundred miles away, although closing in on his location. As they drew closer, he counted nine objects.

What startled me the most at this point was the fact that I could not find any tails on them. I felt sure that, being jets, they had tails, but must be camouflaged in some way so that my eyesight could not perceive them. I knew the Air Force was very artful in the knowledge and use of camouflage. I observed the object's outlines plainly as they flipped and flashed along the snow and also against the sky.[119]

Arnold was a trained observer who had spent thousands of hours in the air and would later take part in Idaho Mountain Search and Rescue flights as well as Mercy Flights efforts, as well as act as a deputy sheriff for the Ada County, Idaho, sheriff's aerial posse.[120] As Arnold started to get a closer look at these objects, he began to realize they were unlike anything he'd ever observed.

I was fascinated by this formation of aircraft. They didn't fly like any aircraft I had ever seen before. In the first place, their echelon formation was backward from that practiced by our Air Force. The elevation of the first craft was greater than that of the last. They flew in a definite formation, but erratically. As I described them at the time, their flight was like speed boats on rough water or similar to the tail of a Chinese kite that I once saw blowing in the wind. Or maybe it would be best to describe their flight characteristics as very similar to a formation of geese, in a rather diagonal chain-like line, as if they were linked together. As I put it to newsmen in Pendleton, Oregon, they flew like "a saucer would if you skipped it across the water."[121]

As Arnold continued to observe the craft, sometime flying together, other times separately, there came a moment when he realized they were traveling in a line between Mount Rainier to the north and Mount Adams to the south.

Checking his watch, he clocked them as traversing this distance in a minute and forty-five seconds. Later, Arnold conservatively estimated the distance between the two peaks to be just under forty miles, which means the craft were going at a speed of at least 1,350 miles per hour.[122] At the time, he believed he must have been seeing some kind of secret guided missiles, for nothing piloted by a human could attain such a speed.

In light of his later interactions with Harold Dahl and Fred Crisman, we think it's important to look at what Arnold wrote of these two very different men. Arnold's story quickly became national news, and as stated earlier, Ray Palmer of *Fate* magazine hired Arnold to investigate a UFO sighting that had happened a few days before Arnold's sighting. After phoning Dahl, Arnold got Dahl to agree to meet him at Arnold's hotel room. Arnold says that when Dahl arrived:

> I invited him in a little apologetically and started firing questions at him. He moved the large chair from near the dresser in between the two beds and let me rave on for a little while, not answering a single thing I asked him. He finally said, "Wait a minute, Mr. Arnold, not so fast. In talking with you there are quite a number of things I want you to consider if you want me to tell my story to you. In fact, I think I better go home." He got up as if to leave, then continued, "Mr. Arnold, I still think it would be good advice to you. This flying saucer business is the most complicated thing you ever got mixed up in."[123]

Although Dahl's words unnerved Arnold, he continued talking to Dahl and finally got him to open up about what was genuinely bothering him. Finally, Dahl partially relented:

> For nearly two hours, Harold Dahl told me of all the sad experiences he had since the 21st of June when he reported his sighting. He said you couldn't blame any of the experiences he had to anyone, but just by coincidence he nearly lost his job, just by coincidence he nearly lost his son, his wife had become ill, that he had lost a tremendously good boom of logs that he had salvaged from the bay when an unusual tide had somehow broken the moorings one night. This was a major loss to his finances as the boom was worth over $3500. The engines on their boat wouldn't start in the mornings; the boat sprang leaks. All in all, he had a horrible time in keeping from going completely broke financially and from losing his family and home through sickness and accident.[124]

Being part of a UFO story rarely seems to be good for a person. Dahl may not have been the first to discover this result, but he certainly is at the beginning of an unfortunate pattern. After complaining about the streak of bad luck, Dahl told Arnold the story of his encounter. As Arnold writes:

[Dahl said,] "As I looked up from the wheel on my boat I noticed six very large doughnut-shaped aircraft. I would judge they were about 2,000 feet above the water and almost directly overhead. At first glance I thought them to be balloons and they seemed to be stationary. However, upon further observance, five of these strange aircraft were circling very slowly around the sixth one which was stationary in the center of the formation. It appeared to me that the center aircraft was in some kind of trouble as it was losing altitude fairly rapidly. The other aircraft stayed at a distance of about two hundred feet above the center one as if they were following the center one down. The center aircraft came to rest almost directly overhead at about five hundred feet above the water.

"All on board our boat were watching these aircraft with a great deal of interest as they apparently had no motors, propellors, or any visible signs of propulsion, and to the best of our hearing, made no sound. In describing the aircraft I would say they were at least one hundred feet in diameter. Each had a hole in the center, approximately twenty-five-feet in diameter. They were all a sort of shell-like gold and silver color."[125]

Dahl went on to describe portholes of approximately five to six feet in diameter, and how finally one of the craft came close to the troubled craft, even appearing to touch it. The troubled craft emitted a dull thud, then began spewing out some form of lightweight metal, which fluttered down like pages of a newspaper, before disgorging a hail of what appeared to be almost like lava rock. This seemed to fix the problem, and the aircraft rose and flew in a westward direction toward the ocean.

Dahl reported this information to Fred Crisman, whom he described as his "superior officer."[126] Crisman said he didn't believe him, but asked for the location and took a boat out the next morning to investigate. Crisman then told Arnold what had happened when he arrived at the site:

"On arriving at Maury Island I did find all the debris, lava rock and some of the white metal that Harold had told me about.

"Now," Fred said, "to make the story worse, while I stood looking at these fragments, holding a few pieces in my hand, one of the same kind of aircraft that Harold described to me came right out of a large cumulus clouds and made a wide circle of this little bay.

"The aircraft was banked at about a ten-degree angle. It had no visible means of propulsion. It was more like a large inner tube with round eyes or portholes around it on the outside. It made no noise and circled the bay as if it was looking for something…. I hold a commercial pilot's license. I flew over a hundred missions in fighter aircraft over Burma and I feel qualified to describe it accurately."[127]

This account Crisman gave Arnold means that Crisman also should be considered an early UFO witness. The alleged physical evidence, the appearance the morning after Dahl's sighting of a mysterious figure warning Dahl not to talk about the sighting, and the crashing of the Air Force plane carrying some suspected UFO materials suggest that something beyond the actions of humans is central to these events.

And yet, Crisman's later alleged involvement in the President Kennedy assassination casts doubt on whether anything Crisman says is reliable. Although the assassination involvement allegation against Crisman was twenty years in the future from the 1947 sighting, it is interesting to read what Arnold wrote of him.

Fred Crisman was a short, stocky fellow, dark-complexioned, a happy-go-lucky appearing person, very cheerful and extremely alert. He was practically bubbling over to tell me his story….

I had the feeling as Crisman talked that, solid as he appeared, he definitely wanted to domineer the conversation and trends of thought about the entire Maury Island incident.

Harold Dahl seemed more in favor of taking little or no part in further discussion of the subject. Dahl in no way attempted to sell his story to me or review it any further.[128]

Even twenty years before New Orleans district attorney Jim Garrison suspected Fred Crisman of being deeply involved in the Kennedy assassination, Kenneth Arnold seemed to have had an uneasy feeling about Crisman.

What's difficult in this series of events is to develop a believable narrative, given Crisman's participation in them.

The first possibility is that Dahl's sighting was part of some elaborate disinformation scheme, although it's difficult to figure out how the downing of an Air Force plane and the death of two military intelligence personnel would be a necessary part of such a plan. How do you get men to choose not to parachute out of a doomed aircraft?

A second is that Crisman's part in these events was something of a coincidence.

A third possibility is that our intelligence agencies had some information about the likely path of UFOs as they were investigating US nuclear facilities in New Mexico. It would make sense to post one of their agents, Fred Crisman, at a potential choke point to observe and manage the story if necessary. If this is true, then this effort was successful—and given his success in shutting down important questions about the phenomenon, Crisman was then able to rise through the newly established CIA, which placed him at a critical position for the assassination of President Kennedy years later.

In the next chapter, we'll look at other early and very important UFO cases, in the hope of ferreting out the roles of our intelligence agencies and any nonhuman beings that may play parts in this decades-old mystery story.

CHAPTER FOUR

LIFE MAGAZINE EXPLAINS UFOS TO THE COUNTRY, THE MEN IN BLACK ATTACK, AND THE HILLS GET TAKEN FOR A RIDE

For those alive in the 1940s, 1950s, or even 1960s, one of the most credible sources of news was the glossy pages of *Life* magazine.

And to the publisher's credit, *Life* published over the years several articles about UFOs, many of which were reprinted in a 2021 book by Kenneth Arnold titled *Life: Report on Flying Saucers*.[129] The first alleged UFO story the magazine covered was the February 1942 "Battle of Los Angeles," in which antiaircraft guns fired for three hours to repeal what the US military believed to be an attack by Japanese airplanes.

However, in July of 1947, *Life* published an article titled, "Speaking of Pictures: A Rash of Flying Disks Breaks Out Over the U.S.,"[130] which gives a relatively fair treatment to the series of sightings, although it also includes some nonserious statements about the phenomenon. After first covering the details of Kenneth Arnold's sighting, the article quickly moves on to other examples.

> In the next 10 days flying disks were reported over 10 states and Canada. Captain E.J. Smith…of United Airlines said he consorted with five disks on the Fourth of July near Boise, blinking his lights at them in the dusk. In Texas, farmer Victor Wenmouth turned from milking his cow Gussie and saw three big, black-bellied disks hovering 300 feet overhead. In Chicago a housewife stood on her porch and watched an airborne object "about the size of a saucer, with legs." She had the feeling that it was about to swoop down and slap her face, so she ran into her house and slammed the door. In Seattle, Wash., Mr.

J. William Sheets said, "Why, they come through our yard all the time." Near Spokane, Wash., Mr. Walter Johnson saw "eight flying washtubs, each about the size of a five-room house. By July 10, disks had been reported in 43 states and the District of Columbia.[131]

Again, we're presented with something of a binary scenario: people were seeing something, or they were not.

While some accounts likely can be dismissed, what are we to make of the reports of trained observers such as Kenneth Arnold and his friend E. J. Smith?

In the April 7, 1952, issue of *Life*, in the upper-right corner next to a picture of Marilyn Monroe leaning seductively against a wall, with text on the left side claiming she's "The Talk of Hollywood," it reads, "There is a Case for Interplanetary Saucers."[132] Sexy blondes and aliens: is there a more American combination? This is how the article "Have We Visitors From Space?" opens:

> For four years, the U.S. public has wondered, worried, or smirked over the strange and insistent tales of eerie objects streaking across American skies. Generally, the tales have provoked only chills or titters—only rarely reflection or analysis.
>
> Last week the U.S. Air Force made known to *LIFE* the following facts:
>
> - As a result of continuing flying saucer reports the Air Force maintains constant intelligence investigation and study of aerial objects.
>
> - A policy of positive action has been adopted to find out, as soon as possible, what is responsible for observations that have been made. As part of this study, military aircraft are alerted to attempt interception, and radar and photographic equipment will be used in an attempt to obtain factual data. If opportunity offers, attempts will be made to recover such unidentified objects.[133]

The article continues with some additional claims, many of which have become a permanent part of UFO folklore. For example, witnesses of high credibility, such as Air Force and commercial pilots as well as scientists and weather observers, were encouraged to report their sightings to the "Air Technical Intelligence Center at Wright-Patterson AFB, Dayton, Ohio" (supposedly where the Roswell artifacts were kept), and all other witnesses were instructed to report their findings to the "nearest Air Force installation."[134] The article also assures witnesses that their reports would be kept in strict confidence and "no one will be ridiculed for making one."[135]

The article then begins a historic recap of what happened since Kenneth Arnold's sighting, noting how quickly UFO reports had come in and the likelihood that many of them were the result of hysteria, but something unusual seemed to be taking place in the skies above America.

> Buried in the heap of hysterical reports were some sobering cases. One was the calamity that befell Air Force Captain Thomas F. Mantell on January 7, 1948. That afternoon Mantell and two other F-51 fighter pilots sighted an object that looked "like an ice-cream cone topped with red" over Goodman Air Force Base at Fort Know, Ky. Mantell followed the strange object up to 20,000 feet and disappeared. Later in the day his body was found in a nearby field, the wreckage of his place scattered for a half mile around...

> Captain Clarence S. Chiles and copilot John B. Whitted were flying in bright moonlight near Montgomery, Ala. [July 24, 1948,] when they suddenly saw "a bright glow" and a "long rocketlike ship" veer past them. They subsequently agreed that it was a "wingless aircraft, 100 feet long, cigar shaped and about twice the diameter of a B-29, with no protruding surfaces, and two rows of windows....

> One night in the summer of 1948 Clyde W. Tombaugh, the discoverer of the planet Pluto, was sitting in the back of his home in Las Cruces, N. Mex. With him were his wife and mother-in-law. It was about 11 p.m. and they were all sitting quietly, admiring the clarity of the southwestern sky, like any proper astronomical family. All at once they saw something rush silently overhead, south to north, too fast for a plane, too slow for a meteor. It seemed to be quite low. All three of the wit-

nesses agreed that the object was definitely a solid "ship" of a kind they had never seen before. It was of an oval shape and "seemed to trail off at the rear into a shapeless luminescence." There was a blue-green glow about the thing. About half a dozen "windows" were clearly visible at the front of the ship and along the side.[136]

Many mainstream media publications, including *Life* magazine, alternately highlighted and discounted the UFO issue, but the above article is critical to understand the thinking of the military in the early 1950s, as well as how the public felt about the issue. The UFOs had been linked to the downing of a military plane, startled two commercial pilots, and seen by arguably the most well-known astronomer of the time, Clyde Tombaugh.

Not only people were observing the UFOs; military radar was tracking them. *Life*'s parade of witnesses continues:

> In this case *LIFE*'s informant is an Air Force officer who holds a top military post at a key atomic base. Since his assignment and whereabouts must be kept a secret, he has asked *LIFE* to withhold his name. He has the highest security rating given. Before he took his present assignment, this officer was in command of the radar equipment that keeps watch over a certain atomic installation. One day in the fall of 1949, while watching a radarscope that covered an area of sky 300 miles wide and 100,000 feet deep, he was startled to detect five apparently metallic objects flying south at tremendous speed and great height. They crossed the 300-mile scope in less than four minutes [a speed of approximately 4,500 miles per hour]. The objects flew the whole time in formation.[137]

While *Life* is careful to note that this could have been an equipment malfunction, the article also reports that "the officer involved is an experienced observer, well aware of the eccentricities of the instrument. He believes in this instance he made a legitimate radar contact."[138] In total, the *Life* magazine article lists ten credible sightings, most with additional witnesses.

Near the end of the article, *Life* quotes Gordon Dean, chairman of the Atomic Energy Commission:

He said: "There's nothing in our shop that could account for these things, and there's nothing going on that I know of that could explain them." Still unconvinced, *LIFE* checked the whereabouts of every scientist who might have anything to do with the development of super-craft. All were accounted for in other ways. Careful feelers through the business and labor world encountered no submerged projects of the immensity necessary to build a fleet of flying disks. And there is still the conclusive fact: U.S. science has at its command no source of power that could put a flying machine through such paces as the saucers perform.

They are not a Russian development [italicized in the original only to denote the things the UFOs are not]. It is inconceivable that the Russians would risk the loss of such a precious military weapon by flying a saucer over enemy territory. No man-made machine is foolproof; sooner or later one would crash in the U.S. and the secret would be out. Nor is there any reason to believe that Russian science, even with German help, has moved beyond not only the practical, but the *theoretical* horizons of U.S. research.[139]

In our modern world, in which we question every source of information, it may be difficult for some to realize there was a time in American history when the vast majority of the public believed what was written in a prestigious magazine such as *Life*. And to a great extent, that public trust was warranted. This was a time when journalists prided themselves on their objectivity, as well as on doing the hard work necessary to track down a story, rather than simply reporting on the opinions of another media personality.

Again, we face a binary scenario.

Life magazine, in conjunction with the military, was doing its very best to make sense of a troubling story.

Or this was the moment when the US military sought to perpetrate what would become a decades-long hoax on the public, for reasons that still remain unclear more than seventy years later.

KENT HECKENLIVELY, JD AND MICHAEL MAZZOLA

▲ ▲ ▲

On August 4, 1952, *Life* published another piece on UFOs, this one titled "Washington's Blips: 'Somethings' over the capital are traced on radar,"[140] and it is even more disturbing than the previous two articles. The opening paragraph sets the scene:

> The most startling "flying saucer" incidents recently reported have taken place during the past two weeks over Washington, D.C. and threaten to make politics take a back seat in the most political of American cities. There, for the first time, mysterious objects in the sky were recorded by ground observers, by pilots in airplanes, and on radar screens all at the same time. And, for the first know time, the U.S. Air Force sent its jet planes up in an attempt to intercept the objects.[141]

The article recounts a story that was gripping the nation at the time. In the early morning hours of July 20, 1952, radar operators started seeing multiple moving blips of light. The movements of those blips were unlike those of any known craft. Sometimes they'd idly remain in one location, at other times they'd move forward and then quickly reverse direction, and at still other times, they'd rocket across the sky and then take a left turn at a ninety-degree angle.

Observers at Andrews Air Force Base and Washington National Airport observed the lights, and at one point, three radars independently verified a craft. The article includes the observations of Harry Barnes, a senior air traffic controller at Washington airport.

> Although Barnes has estimated that some of the objects dawdled along as slowly as 130 mph, others went so fast that his radar could not track them. However, the radars at the airport towers, apparently capable of tracking faster-moving bodies, were able to fix on one object long enough to show that it had traveled eight miles in four seconds, which meant that its speed was 7,200 mph.

> It was not until 3:00 a.m., two hours after Barnes's call, that radar-equipped jet fighters roared in from their Delaware base and called Barnes by radio. They reported they saw nothing.

Barnes agreed that there were no unidentified targets on his scope at that moment. The planes, low on fuel, returned to their base. Shortly afterwards the blips were erupting all over the radar scope again.[142]

One gets the sense that these craft were toying with the radar operators, and the radar operators were dealing with a novel situation. Again, we're presented with a remarkable range of maneuvers by these objects, including having the ability to hover, to accelerate to remarkable speeds, and to make turns at velocities that would kill a human pilot.

The following Saturday night, starting at 9:08 p.m., can only be described as some kind of strange duel between American aircraft and the unexpected visitors. At the end of the article, the *Life* writer rebukes the Air Force for its lack of transparency.

...Barnes had already notified the Pentagon Command Post, the high brass in Washington, and the Air Defense Command. From their Delaware base, F-94 jet interceptors again barreled down toward Washington. They arrived at 11:25 and howled over the city. What happened then is in dispute. Officials in the radar room firmly state that a pilot reported contact at 11:25 with four lights 10 miles away and 500 feet above him. He closed at full throttle for two minutes, but the lights disappeared at tremendous speed. Another contact was made a few minutes later and was similarly broken off.... But when questioned by LIFE the pilots themselves denied any certain visual contacts with aerial lights or objects.

The attitude of the Air Force during the July incident was puzzling. When the first appearance of the blips were reported in Washington newspapers, no mention was made of jet interceptors. In fact, the Air Force stated that it had sent none up. But when confronted with the facts by TIME-LIFE Washington correspondent Clay Blair, Jr. who gathered the material for this article the Air Force finally admitted it had indeed sent the fighters up.[143]

It's understandable that some would claim this was the beginning of a decades-long hoax by the military and intelligence agencies. It seems completely plausible that those responding on behalf of the Air Force were genuinely panicked over what they saw, and decided the best thing to do was to calm the public while they investigated.

Perhaps there are things in the night of which we should be afraid. But it's not a good strategy to pretend they don't exist.

▲ ▲ ▲

A week later, on August 11, 1952, *Life* reported on a press conference by Air Force lieutenant general John Samford, in which he essentially told the public to calm down about flying saucers.

> Chief of Intelligence, Major General John Samford said the radar "blips" might be due to an atmospheric condition known as "inversion" a layer of cool air between layers of warm air which can, under certain conditions, reflect radar rays. Although usually only solid objects appear on scopes, radar mirages were familiar during the war to Navy ships, which occasionally fired at nonexistent "objects." The visual sightings confirming the radar reports might, the Air Force said, be reflections from light sources on the ground. At any rate people could stop worrying about flying saucers.[144]

Our response to this article may be unexpected.

From a military point of view, we understand why Samford may have tried to calm the public about UFOs.

Just for the sake of argument, let's say everything we've written in this book so far about the early history of UFOs is true.

Let's accept as true that a month after the United States exploded the world's first atomic bomb in July of 1945, a UFO crashed near the Trinity test site, where it was discovered by two young boys on horseback.

Let's accept that on June 21, 1947, Harold Dahl did observe several disc-shaped craft, including one that was in trouble and had to be assisted by other craft to avoid crashing. In the process of repairing itself, the craft discharged a large amount of material that damaged Dahl's boat, injured his son, and killed his dog.

Let's accept that on June 24, 1947, pilot Kenneth Arnold also saw several strange aircraft traveling at fantastic speeds, with remarkable maneuverability, in the skies above Washington state.

Let's accept that on July 1, 1947, a plane carrying alleged UFO materials from Harold Dahl's encounter mysteriously crashed, killing two military intelligence officials, who should have been able to safely parachute out of the troubled aircraft.

Let's accept that sometime during the night of either July 3 or 4, 1947, a UFO crashed in Roswell, New Mexico, and that it was first reported by the national news as a downed saucer, which the military claimed to have been a weather balloon.

Let's accept that in the ensuing five years, from 1947 to 1952, many credible observers made flying saucer reports. These included military and commercial pilots as well as the most famed astronomer of the time, Clyde Tombaugh, discoverer of Pluto.

And finally, let's accept that in the summer of 1952, these objects were buzzing though the most secure airspace in our country, that of Washington, DC, the very seat of our government. Even when we sent up our most advanced fighters, these objects played with them like a cat might with a terrified mouse.

If I was a patriotic military man, what would be going through my mind?

The first thing I'd realize is that I don't know what's going on, but the evidence at hand suggests some broad outlines of the capabilities of this possible foe.

While the capabilities of these craft are remarkable, they also seem to be somewhat fragile in Earth's environment. Consider the Trinity crash in August of 1945; the ship in distress Harold Dahl saw in June of 1947 which was assisted by its fellow craft; and the Roswell crash of July 1947. This fragility possibility would be reinforced by the observation that these craft seem to travel in groups, most likely because they understand that there are likely to be equipment malfunctions that the others might need to fix.

By 1952, there would have been more than enough time to examine the bodies, and to likely conclude that these creatures were engineered and we do not know their true identity. The "mysterious man" Harold Dahl reported would raise suspicion that these visitors have the ability to pass as human, and subsequently that they have long been observing us and yet keeping themselves hidden for some reason.

The most likely historic parallel we draw is the European contact with Native Americans. Yes, the ships on which the Europeans traveled would fill the native people with awe. But the reality was that the Europeans were remarkably vulnerable. If the natives had adopted a policy of exterminating such outsiders upon their first appearance, there is little chance that the Europeans could have established a foothold in the Americas on the time-table they did. The Europeans had to appear peaceful to the native people, hence giving them gifts, and be a benefit to any group that befriended them.

In other words, the Europeans had to lie to the Native Americans about their ultimate aims in order to establish a foothold in the New World. Could the Wampanoag tribe that kept the Plymouth Rock colony alive that first brutal winter have imagined that these people who seemed so unsuited to their new environment would conquer the continent within a few hundred years?

If I were a military man assessing this threat, I'd ask myself why the aliens didn't simply announce themselves at an earlier point in history and share with us what they knew. Can you imagine how different history might have been if at some time early in the Industrial Revolution of the nineteenth century, these aliens had simply made themselves known? If their intentions were benevolent, couldn't they have dramatically changed our history? We would not have had to pollute so much of our planet, and we could have avoided the rampant poverty and slaughter of so many in nationalistic wars.

We believe the revelation of the existence of aliens would have led to a lessening of man's violent nature, not a disintegration of society.

But the aliens did not choose to make themselves known.

If we were military men, we might conclude that our welfare likely was not the most important thing to the aliens. What we would know is that these beings had an advantage over us, and if nature is any guide, that usually never ends well for those holding the short end of the stick.

▲   ▲   ▲

While we're comfortable quoting liberally from *Time* magazine because of its well-established credibility for truth (at least during that time), we're less confident about our next source, a curious book titled *Flying Saucers and the Three Men*, by Albert K. Bender, first published in 1962. The book was recommended to us because it's supposedly the first time in print that the expression "men in black" was used, and much of the later mythology around these figures springs from this initial account.

Albert Bender was certainly a curious character. He lived with his stepfather in a large room on the third floor of their house, adjacent to a large attic. He loved the noises the house would make at night, imagining they were ghosts. He also enjoyed horror and science fiction stories. As he describes it:

> Having read many weird stories written by such greats as Mary Shelley, Bram Stoker, Edgar Allen Poe, and Edgar Rice Burroughs, I decided to depict some of the characters they had invented by painting grotesque scenes and faces upon the walls of the room. I began the task in earnest fashion, and after about eight months I had done so good a job that it almost frightened me when I stood back and looked at it all one evening. No wonder my friends found it so fascinating, for so many of the ghostly characters appeared to be looking straight at me, no matter where I might be in the room![145]

In many ways, Bender does not seem to be your normal guy. And yet, he had a regular job, had friends, and eventually got married. As reports continued in the country about "flying saucers," Bender became interested in the subject. In April of 1952 (around the time of the *Life* article that detailed the US Air Force's newfound openness to UFOs and included a list of impressive sightings), he founded the International Flying Saucer Bureau (IFSB).[146] It quickly became popular, and he started publishing a quarterly newsletter called *Space Review*.[147] Soon the venture had branches in England, France, and Australia, and the effort seemed quite successful.

> During January and February of 1953, so many sightings and reports were arriving at IFSB headquarters that we called a meeting of the executive staff to discuss means of handling the information more efficiently, and to find better ways of

checking it for authenticity. The staff decided a Department of Investigation should be formed. This, we felt, should consist of a chairman or head, with four other people to work with him. The executive staff could forward the most important sightings to this Department of Investigation and possible firm conclusions.[148]

The IFSB staff had three issues under their belt in 1953, and many members were joining, interested in what was clearly a matter of public interest, especially since the UFO sightings over Washington, DC, had been reported. To coordinate this effort, Bender hired Gray Barker, who had done a good job of investigating an earlier UFO case in West Virginia, which became known as the Flatwoods Monster incident; several witnesses reported seeing a ten-foot-tall robot-looking creature, as well as unidentified objects. Barker received business cards when he accepted the new position, but that brought him an unwelcome visitor. As Bender writes:

A few days after he received the cards an FBI agent called at his office, with one of the cards in his hand. Barker told me this unnerved him because he immediately felt that there must be something wrong with our organization and that he was in trouble as a result. The agent asked what Barker knew about the organization, then relieved his fears by explaining the reason for the visit. He obtained the card from a man who had suffered an epileptic fit and was a patient in a local hospital. On the back of the card was written a name and a Florida address, but Barker had been too flustered to write it down.

The agent departed after a few more questions, and Barker was left with a puzzling mystery. He had just received the cards a few days before, had given out none of them except to four or five close friends, and one hitchhiker. None of his friends were epileptics, and the hitchhiker was a serviceman headed for somewhere in the South.[149]

Was this a genuine FBI agent, or somebody from a little farther out of town? The story of Barker's business card's being found with a man who'd suffered an epileptic seizure didn't make a good deal of sense. Bender was unnerved by the report and wished Barker had gotten more information, but stranger events would soon take place. The members of Bender's organization were also concerned about the Air Force and their intentions.

A subject of friendly controversy at many local IFSB meetings was the U.S. Air Force policy on flying saucers. Some of the executive staff agreed with Major Donald E. Keyhoe who, in his writings and lectures, charged that the Air Force was deliberately withholding certain information which might prove the saucers to be of interplanetary origin. Other staff members, along with myself, weren't certain, and we argued that the Air Force might not know a great deal more than did the IFSB, even though they had incomparably better facilities to carry out their investigations.[150]

It can be shocking to read accounts from the 1950s that make it seem as if people were just as confused about UFOs then as we are in the 2020s. The validity of the story of Barker's visitor, and its mysteries, may be one of its most defining factors.

And the idea of an Air Force whistleblower saying the government was concealing vital information sounds as if it might be ripped from a modern-day headline, although this instance took place in 1952.

Bender's book highlights many of the cases his organization investigated, but then came the incident that led him to shut down the IFSB. The bureau was planning to hold an event called World Contact Day, on March 15, 1953, at several locations in the United States, England, France, Australia, New Zealand, and Canada, in which the participants would try to mentally communicate with the aliens and encourage them to reveal themselves.

But Bender got an unexpected response. As he describes in his book:

I opened my eyes, and to my amazement I seemed to be floating above my bed, but looking down upon it where I imagined I could see my own body lying there! It was as if my soul had left my body and I was hovering above it about three feet in mid-air. Suddenly I could hear a voice, which permeated me but, in some way, did not seem to be an audible sound. The voice seemed to come from the room in front of me, which remained pitch dark.

"We have been watching you and your activities. Please be advised to discontinue delving into the mysteries of the universe. We will make an appearance if you disobey."

I replied in words, though my lips did not move: "Why aren't you friendly to us, as we do not mean to do any harm to you?"

"We have a special assignment," came the reply, "and must not be disturbed by your people."

As I tried to remonstrate, I was interrupted by another statement: "We are among you and know your every move, so please be advised we are here on your Earth."[151]

Bender discounted this curious occurrence, chalking it up to being a dream, perhaps brought about by concern over Barker's visit from an alleged FBI agent as well as some curious interviews an official of the Australian branch of the organization had had with that country's law enforcement officials. Bender told some of his friends about the experience, but they responded negatively, suggesting he was either lying or had gone crazy. He tried to put the experience out of his mind and continue as before, but that brought the threatened visitation. As he writes:

The room seemed to grow dark yet I could still see. I noted three shadowy figures in the room. They floated about a foot off the floor. My temples throbbed and my body grew light. I had the feeling of being washed clean. The three figures became clearer. All of them were dressed in black clothes. They looked like clergymen, but wore hats similar to Homburg style. The faces were not clearly discernible, for the hats partly hid and shaded them. Feelings of fear left me, as if some peculiar remedy had made my entire body immune to fright.

The eyes of all three figures suddenly lit up like flashlight bulbs, and all these were focused upon me. They seemed to burn into my very soul as the pains above my eyes became almost unbearable. It was then I sensed that they were conveying a message to me by telepathy.[152]

The science fiction writer Arthur C. Clarke is known for his famous saying, "Any sufficiently advanced technology is indistinguishable from magic." Let's assume that aliens have a technology that mimics telepathy, allowing them to instill images and messages in the minds of the target individuals, perhaps in a way that Elon Musk is currently attempting to do with his Neuralink devices.

If, like the early European explorers, you were traveling to a new land, uncertain of how the native population would receive you, this would be an enormously useful tool. It would be almost like a tranquilizer gun used to incapacitate large game animals you wish to tag, move to a different location, or take biological samples from. We find it interesting that Bender describes it in this way: "Feelings of fear left me, as if some peculiar remedy had made my entire body immune to fright." With Bender in such a calm state, the aliens felt they could impart their message to him.

> You are not a person of great renown on your planet; therefore we have nothing to fear at present. We have a purpose for being here and we will be here for some time yet. We must not be disturbed in our ultimate goal. As you see us here, we are not in our natural form. We have found it necessary to take on the look of your people while we are here. This is mainly used as a means of returning here without being detected by anyone. We have made numerous contacts with Earth by means of craft from our own base, and at present we have craft hidden at a remote spot on your planet. We have found it necessary to go to great extremes at times to frighten off your Earth people, and it has resulted in their deaths.
>
> We also found it necessary to carry off Earth people to use their bodies to disguise our own. We wish to keep in touch with you, and tell you many things, because one day you will write about this, and we are certain that nobody will believe you, but you will be much wiser than anyone else on your planet. You will know what is out there in space, and you will know what the future holds for mankind.[153]

We remind the reader that Bender's book was published in 1962, and yet it reads like a template of many more recent UFO encounters. Prior to this experience, Bender was of the opinion that these visitors could be best compared to "Space Brothers" (a term coined in the late 1940s and early 1950s by George Adamski, who claimed to have met with benevolent Nordic looking aliens), who had the best interests of humanity at heart.

The reality might be something significantly different.

If Bender's account is to be believed, the aliens don't want us messing with their plans and are willing to kill those who continue to pursue the truth.

The aliens also don't seem to have any qualms about kidnapping humans to use for their own purposes.

If these aliens are supposed to be our space brothers, I'm not sure that's a family of which I want to be a part.

According to Bender, the aliens continued instilling their messages, telling him that their plans would continue for a number of years[,] but that they would give him a piece of metal that would disappear when he was free to talk about his experiences.

In another communication from these "men in black," they explained that there are numerous alien civilizations, but that at any one point in time, only a few of these are able to reach Earth.

> ...The nearest planet to Earth at one time nurtured a great civilization, which was destroyed by marauders from another system of planets in an orbit beyond ours. They will once again make their appearance in the future when they reach this same spot in their trip around the central body. Almost every system of planets that has an orbit around the central body contains some sort of intelligent inhabitants, but not all the same in body structure, being adapted to the various conditions that exist on their particular worlds. Because many of these are far advanced in their ways of life, your planet Earth will be constantly under surveillance by these systems as they overtake you and pass you by. Your planet is yet an infant as far as progress is concerned, and you have far to go to accomplish what many others in your neighboring systems have already achieved.[154]

How scary are these inhabitants and systems? Well, according to this alien race, there was another alien race that destroyed life on Mars. Maybe spending a couple of trillion dollars on weapons to defend ourselves might not be such a bad use of resources.

But at least we can rest assured that the aliens who allegedly visited Bender didn't have the capacity for such violence, right? I'll let Bender tell you in his own words what his first visitor telepathically conveyed to him.

He then paused in this discourse, and with a motion of his hand changed the picture on the tube. I saw a landmark familiar to every American: the Pentagon in Washington, D.C., and the surrounding areas.

"You wish to know why I am showing you this view," he continued. "It is only to inform you that we have some of our people stationed in your so-called Pentagon while we are visiting your planet. We have them stationed in numerous places around your planet, to keep us informed of all that is taking place."[155]

Bender's account suggests not only that these aliens are able to assume human form to clean up any problems associated with their being seen by people, but that they have several individuals prepositioned to monitor and quickly intervene as necessary. However, it got much worse.

The next scene showed a vital spot in the United States, but unidentified as to locale. It was one of our atomic stockpiles. Then continued changes of scene exhibited similar storage places in the United States and other countries, including the Soviet Union. I asked him why he was showing this to me, and he startled me with his blunt answer. I sensed he wished to appear friendly, but his reply led me to doubt.

"With the push of a small button in our space laboratory we can detonate every bomb you have in your stockpiles all over the globe, causing almost total destruction of your planet."[156]

When Albert Bender started his UFO investigation, this was probably the last thing he expected to hear from the representative of a supposedly advanced civilization. They might possess the ability for interstellar travel, but their moral evolution didn't appear to be any more advanced than that of a genocidal tyrant.

Power, not justice, seemed to be their operating principle.

Bender struggled to comprehend this information. As he says:

The first thing that came to my mind was the question, "Why would you want to do something so horrible?"

His reply was blunt and to the point: "Only if we were discovered and your people tried to stop us with whatever means they had at their disposal. But having looked over your planet thoroughly we have nothing to fear in this respect, for nothing you have on Earth could harm us. Our weapons for self-defense against marauders in space are far superior to anything you have."[157]

There's a great deal to take in from that passage. If true, these aliens have several of their number positioned at places on Earth to observe what we're doing. And if we do the wrong thing, they'll blow us to kingdom come. The being went on to tell Bender that he and his group were on a several-years-long mission to extract large amounts of a rare element from our seawater, using their hidden base in Antarctica.

And yet we can't help but notice that there are a couple of inconsistencies in what Bender reports. If these aliens have weapons that can defeat the space marauders who laid waste to Mars, why would they need to booby-trap our own nuclear bombs?

If we have been under their surveillance for a long time, it appears that something about humans worries them, as when the being states, "Only if we were discovered and your people tried to stop us with whatever means they had at their disposal."

This statement suggests to us that *we could stop them.*

Perhaps among the galactic races, we humans are particularly good innovators, able to quickly assess new information and figure out how to overcome any disadvantage.

Of course, on the negative side, the aliens might destroy our world as they did Mars.

Bender's account leaves me with more questions than answers.

If the aliens had assessed Earth's defenses and concluded that all of our military might was unlikely to pose a threat to their efforts, why waste any time on a little group like Bender's International Flying Saucer Bureau?

Were they simply bored and looking to play an elaborate practical joke to pass the time?

The actions of the aliens, from the taunting of our military aircraft and violation of restricted airspace, as well as casually mentioning to Bender that they'd booby-trapped all of our nuclear stockpiles, strikes us as suggesting not power, but weakness.

More than fifteen hundred years ago, in *The Art of War*, Chinese tactician Sun Tzu famously wrote that "all warfare is based on deception," and that "when one is weak, they should appear strong." The earliest accounts of alien encounters suggest that at least some of the aliens were struggling with keeping their craft from crashing on Earth, especially between 1945 and 1947.

By 1952, it appeared that they had made some modifications and were brazen enough to challenge our jet fighters over Washington, DC.

But did that mean that their underlying weakness, possibly the difficulty in getting to our planet or moving with ease through it, had substantially changed?

Most of the speculation about aliens concerns whether they, like some science fiction monster, wish to wipe us out, or whether we should consider them our spiritually advanced space brothers, almost like angels of the cosmos, encouraging us on a more peaceful path.

What if reality is a little more complicated? What if in the aliens' intentions, there's some good, some bad, and a lot of stuff that doesn't fit neatly into either category?

Maybe the aliens fear that if they were to reveal themselves, to let us poke and prod them as they have obviously done with us, we'd figure out them and their technology.

Maybe the aliens don't want us to realize that we could be their equal.

A A A

Albert Bender's story was the first to popularize the idea of the "men in black," although there may be earlier examples.

In much the same way, the Betty and Barney Hill story has become the template for the missing time/alien abduction story we have come to know so well through books, movies, and television shows. The Hills allegedly were abducted by/made contact with aliens from September 19 to September 20, 1961, while driving in Exeter, New Hampshire.

Because the 1965 book about their story, *The Interrupted Journey: Two Lost Hours Aboard a Flying Saucer*, is the first significant UFO abduction case published, and because it involves the use of hypnotic regression to uncover their alleged memories of it, it's important to closely examine this case.

As to what was happening during that time in the area, I quote from the foreword to the book, written by the author, a local newspaper reporter at the time of the incident, John G. Fuller. The alleged abduction of the Hills took place against the backdrop of a rash of UFO sightings.

> My research in the Exeter area extended for several weeks. I had at first suspected that the UFO story could be explained by careful, painstaking research in a single area, and that a rational answer should turn up. It didn't. Police, Air Force pilots and radarmen, Navy personnel, and coastguardsmen all confirmed the incredible reports that dozens of reliable and competent citizens in the area were giving me in grueling cross-examinations.

> I took advantage of the Exeter police station as a base from which to conduct the research, since current reports of the phenomenon gravitated there. Toward the end of my research period, a message was left at the police station that Mr. and Mrs. Hill would appreciate it if I'd call them in nearby Portsmouth.[158]

What followed was confusing for a period of time, because while Betty and Barney Hill recalled seeing a UFO, there were gaps in their memories of the night. They could recall pulling their car over and seeing a craft coming down with illuminated windows and beings inside, and how they then jumped in their car and sped away. However, their memories seem to be missing a few hours, from the time they fled the spaceship until they were driving down the road several miles away. This is how Fuller describes the couple:

> ...Barney, a strikingly handsome descendant of a proud Ethiopian freeman whose great-grandmother was born during slavery, but raised in the house of the plantation owner because she was his own daughter; Betty, whose family bought three tracts of land in York, Maine, only to have one member cut down by Indians. Regardless of what attention their mixed marriage drew in public places, they were no longer self-conscious about it. Their first attraction to each other, one that still remained, was of intellect and mutual interests. Together, they stumped the state of New Hampshire, speaking for the cause of civil

rights. Barney, former political action, and now legal redress chairman of the Portsmouth NAACP was also a member of the State Advisory Board of the United States Civil Rights Commission and the Board of Directors of the Rockingham County Poverty Program. Both he and his wife are proud to display the award he received from Sargent Shriver for his work. Betty, a social worker for the state of New Hampshire, continues after hours with her job as assistant secretary and community coordinator for the NAACP, and as United Nations envoy for the Unitarian-Universalist Church to which they belong in Portsmouth.[159]

Betty and Barney Hill were clearly credible people, pillars of their community, commendable for their interest in civil rights. The part of their memories that was clear prior to the hypnosis sessions was a driving trip from their home in Portsmouth, New Hampshire, to Niagara Falls, then to Montreal, Canada, and back home. On their return trip home, four nights later, just south of Lancaster, New Hampshire, they saw an unfamiliar object, which at first they took to be a search plane because it was flying so low.

But as they continued their drive, the object seemed to be following them.

They stopped at a picnic turnout with an overlook and got out their binoculars to get a better look at the object, but couldn't decide what it was and continued on their journey. As the craft continued to follow them, getting even closer, Betty told Barney he should stop the car so he could see it through the binoculars. Barney estimated that the craft was no more than two hundred feet in the air but keeping a safe distance from them.

> Barney stopped the car almost in the center of the road, forgetting in the excitement any problems with other traffic. "All right, give me the binoculars," he said. Betty resented his tone. It sounded as if he was trying to humor her.
>
> Barney got out, the motor still running, and leaned his arm on the door of the car. By now the object had swung toward them and hovered silently in the air not more than a short city block away, not more than two treetops high. It was raked on

an angle, and its full shape was apparent for the first time: that of a large glowing pancake....

"Do you see it? Do you see it?" Betty said. For the first time her voice was rising in emotion. Barney, he frankly admitted later, perhaps as much because Betty rarely became excited as because of the nearness of this strange and utterly silent object defying almost any law of aerodynamics.[160]

Although Betty started yelling at Barney to come back to the car, he continued moving towards the craft.

Barney was fully gripped with fear now, but for a reason that he cannot yet explain, he found himself moving across the road on the driver's side of the car, onto the field and across the field, directly toward him. Now the enormous disc was raked on an angle toward him. Two finlike projections on either side were now sliding out farther, each with a red light on it. The windows curved around the craft, around the perimeter of the thick, pancake-like disc, glowing with brilliant white light. There was still no sound. Shaken, but still finding an irresistible impulse to move closer to the craft, he continued on across the field, coming within fifty feet of it, as it dropped down to the height of a single tall tree. He did not estimate its size in feet, except that he knew it was as big or bigger in diameter than the length of a jet airliner.[161]

Again, if we take a skeptical look at the story, these aliens seem to have some technology that overcomes the will of those in close proximity. Barney claims he didn't hear Betty yelling at him, which under these circumstances makes it seem as if the aliens were in control of his responses. He also could have been in shock.

It's reasonable to question whether Barney was simply so mesmerized by the craft that he acted of his own volition. But based on what transpired later, we should not discount the possibility that the aliens had effectively taken control of Barney's actions.

Eventually, Barney did gain control of himself, raced back to the car, and drove away. But the craft continued its pursuit, and he and Betty barreled down the road, trying to make their escape, they heard "in irregular rhythm—*beep, beep—beep, beep, beep*—seeming to come from behind the car," and "[f]rom that moment, a sort of haze came over them."[162]

While they recalled part of their experience, they knew there were some missing hours. As the *Life* magazine article from 1952 had instructed witnesses to do, the Hills reported their experience to Pease Air Force Base, the nearest Air Force facility. Their report was eventually forwarded to the main UFO reporting facility, Wright-Patterson Air Force Base in Ohio, and included in Project Blue Book, an early attempt to synthesize UFO sightings.

Betty and Barney reacted differently to the situation. Betty was more curious, reading the book *The Flying Saucer Conspiracy*, by Donald E. Keyhoe, which claimed the Air Force was not making a serious effort to study the phenomenon. Barney struggled, trying to separate fact from fantasy. Betty began to have terrifying dreams of being abducted by strange men, and once, when she and Barney were driving and became momentarily delayed by a car partially blocking the road, Betty had something of a panic attack and urged Barney to drive quickly away.

The report of the Hills attracted the attention of one Air Force official:

> As a hardhearted former intelligence officer in the Air Force, Major James McDonald groped for some kind of answer to the puzzle. UFO's are constantly being discussed in the Air Force, much more so than the laconic official statements from the Pentagon indicate. Officially, the Air Force position requires that no member of the force can report any incidents to the public; all information must be channeled through the Foreign technology Division at Wright-Patterson Air Force Base, Ohio, and in turn, any release of that information can be made by the Office of the Secretary of the Air Force at the Pentagon. The fact remains that many Air Force pilots and radarmen do talk, and those who have directly come into contact with the objects reveal stories of incredible speeds, right-angle turns, and maneuvers that are impossible to duplicate by any aircraft known to the military. Even the most sophisticated weapons were said to have been used in an attempt to bring down the UFOs without success.[163]

Some commentators have been suspicious of MacDonald, given his background as an intelligence officer, and the fact that it was MacDonald who "suggested the possibility of medical hypnosis"[164] for the Hills.

It was this therapy, hypnotic regression, that would reveal the most shocking parts of the alleged Hill abduction story.

▲ ▲ ▲

Hypnotic regression has been criticized as being unreliable, especially given the well-established fact that one of the dangers is that a practitioner, even innocently, may unknowingly plant false memories in the subject.

However, others argue that hypnotic regression can be of great value for those who have suffered traumatic events, which they may have trouble processing. In the case of the Hills, it must be noted that even prior to hypnotherapy, they knew "something" had happened that night in 1961, and there were extensive occurrences, such as seeing the craft, that they could recall without resorting to hypnosis.

In these sessions, Barney recalled a great deal of what had already been known, such as the details of their trip to Montreal (having to deal with a racist motel manager who wouldn't rent to an interracial couple and having to find a different hotel) and their initial observations of the craft on their drive home. In the first session, psychiatrist Benjamin Simon asked about the couple's prior belief in UFOs. Barney recounted how Betty's sister had talked about seeing a UFO years before, which had made Betty more open to the possibility. From the transcript:

DR. SIMON: Did she have any reason for believing in them?

BARNEY: Her sister. I am thinking of... [a visit I had with] her mother and father in Kingston, New Hampshire. And they live in a nice, quiet area. Only three houses—her two sisters' and her mother's houses are located there. And at night you can look at the sky and see millions of stars. And I think how beautiful this is. And we were talking about satellites. The Russians had sent up Sputnik. And her father was talking about, and how you could see some satellites from here at certain hours. And we talked about flying, we talked about life on other planets. And then Betty's sister said she had seen an object flying, long and cigar shaped, and smaller objects coming to it and flying away from it.[165]

There's no doubt that Barney had thought about UFOs, but the visit to Kingston had taken place four years earlier, and in his session, stated he hadn't thought much about it in the intervening years. What is curious to me from the account of Betty's sister is the observation of two types of ships—a large, cigar-shaped craft and a smaller, presumably disc-shaped craft, flying away from it. Similar observations by others would lead to claims that these smaller craft were functioning as "scouts" much like our current drone technology.

At one point in his session, reliving the event, Barney shouted, "I gotta get a weapon!"[166] The doctor calmed him down and told him he was safe now and could proceed with his story.

Barney continued the session, which was punctuated by occasional moans and cries of distress. As he got closer to the craft, reliving it in his mind, he calmed down. He looked at the UFO through his binoculars and saw two figures, one of whom reminded him of an Irishman, and the other of a Nazi officer, although the figures had slanted, Asian eyes.[167] As he was observing these figures, Barney had the sense he was being hunted like a rabbit.[168] In his mind, he tore himself away from the scene and fled with Betty in their car.

What happened next is unclear from the transcript, except that Barney said the craft continued to pursue them, and that he had the sensation of being separated from his body.[169] The therapist was concerned about Barney's obvious discomfort and ended the session.

▲   ▲   ▲

At the next session, they were able to go much deeper. Barney began by telling the doctor that after their last session, he'd begun to have troubling memories of eyes, almost like those of the Cheshire Cat from *Alice in Wonderland,* looking at him. Then the two returned to where they'd left off. Barney described stopping the car after his and Betty's initial encounter, because there seemed to have been some sort of accident. He thought of the tire wrench in the trunk as a possible weapon.

BARNEY: I saw a group of men, and they were standing in the highway. And it was brightly lit up, as if it were almost daylight, but not really day....

And they began coming toward me. And I did not think after that of my tire wrench. And I became afraid if I did think of this as a weapon, I would be harmed. And they came and assisted me....

I felt very weak. I felt very weak but wasn't afraid. And I can't even think of being afraid. I am not bewildered, I can't even think of questioning what is happing to me....

[Note that Keyhoe added in The Flying Saucer Conspiracy:] ...Later, when Barney listened to the playback of the tapes, he likened this event to the feeling he had when he went into hypnosis with the doctor. The questions have since been on his mind: If this is true, was he being put into hypnosis by these "men," and if so, was his amnesia caused by this?[170]

The questions Barney Hill asked in 1964 are just as relevant and mysterious today as they were sixty years ago. Could there be a more accurate description than of these aliens hunting us like rabbits and maybe just trying to make sure our hearts don't go pitter-patter too fast, causing us to injure ourselves?

It seems to us that if there was some sort of authentic telepathy, and one was convinced that these aliens meant no harm, there would be no later trauma associated with the event. One may be momentarily surprised by walking into a dark room, having the lights come on, and hearing a crowd of twenty people shout, "Surprise! Happy birthday!", but our later recollection is not filled with trauma.

From Betty's hypnotic regression session, there's a curious encounter that doesn't seem to make a great deal of sense, and it has to do with the aliens explaining where they were from:

...And he went across the room to the head of the table and he did something, he opened up, it wasn't like a drawer, he sort of did something, and the metal of the wall, there was an opening. And he pulled out a map, and he asked me if I had ever seen a map like this before. And I walked across the room and I leaned against the table. And I looked at it. And it was a map…. And there was one big circle, and it had a lot of lines coming out from it. A lot of lines going to another circle quite close but not as big. And these were heavy lines. And I asked him what they meant. And he said that the heavy lines were trade routes. And then the other lines, the other lines, the solid lines were places they went occasionally. And he said the broken lines were expeditions….

So I asked him where was his home port, and he said, "Where are you on the map?" I looked and laughed and said, "I don't know." So he said, "If you don't know where you are, then there isn't any point in my telling you where I am from."[171]

These aliens didn't appear to be very helpful to humans, whom they certainly knew would be terrified by the abduction. Let's just play the substitution game and imagine if humans had gone to an alien world and acted in such a manner. There are a few possible explanations for Betty and Barney's story; first, this is all some elaborate hoax; second, it's a psychotic episode improbably affecting two people; or third, it's real, and these visitors don't seem to have the best of intentions.

The discussion Betty had with the apparent leader of the group was interrupted by some other beings coming to talk about an incident with Barney, who apparently was in a nearby room. Betty was worried that something had happened to Barney, but it was apparently not a serious problem.

BETTY: …The examiner said they could not figure it out—Barney's teeth came out, and mine didn't. I was really laughing and said Barney had dentures, and I didn't, and that is why his teeth came out. So then they asked me, "What are dentures?" And I said people as they got older lost their teeth. They had to go to a dentist and have their teeth extracted. And the leader said, "Well, does this happen to many people?" He was—uh—he acted as if he didn't believe me. And I said, "Yes, it happens to almost everyone as they get older." And he said, "Well—older. What is older?" I said, "Old age." So he said, "What is old age?"[172]

Could Betty and Barney Hill have been kidnapped by the dumbest aliens in the entire galaxy? They didn't seem like they were the A-Team in terms of intelligence, and didn't have even a cursory knowledge of Earth.

Another possibility is that the beings that pilot the smaller, saucer-like craft, are genetically engineered idiots, not able to give any useful information to humans in the event that they are captured and interrogated. At least that's a better explanation to us than beings who don't know about dentures or old age.

But we don't think the aliens who created the craft, or the beings that fly the smaller ships, are that stupid.

We think they're just as wily as the Europeans who crossed the Atlantic in small ships, often enduring conditions of near starvation, then landed on the shores of New World and knew that if they didn't make nice with the locals, there was no way they were going to establish those first colonies.

Given that aliens have apparently been studying us and working missions for a long time, perhaps it's time for us to start coordinating our efforts to learn more about them. It doesn't seem like their idea of sharing is any better than in the story of the Europeans in the New World.

And we certainly wouldn't want to end up like the Native Americans in that latter scenario.

CHAPTER FIVE

THE CLAIMS OF THE FORMER SPOKESWOMAN FOR WERNHER VON BRAUN, THE NAZI ROCKET SCIENTIST WHO GOT AMERICA TO THE MOON

The name Carol Rosin should be more familiar to Americans than it is. In the early 1970s, Rosin became the first female corporate manager of an aerospace company, Fairchild Industries. She was a space and missile consultant who consulted with various defense contractors, government agencies, and the intelligence community. She was a consultant to TRW (Thompson Ramo Woolridge, an aerospace contractor, best known for building the first intercontinental missile, the Atlas rocket, and Pioneer spacecraft) on the MX missile, the marketing of which turned out to be a model for how to sell space-based weapons to the public.[173]

In early 1974, as a corporate manager at Fairchild Industries, Rosin got the opportunity to meet Wernher von Braun, who at the time was probably the most famous scientist in America, having led the Apollo program that landed Neil Armstrong and Buzz Aldrin on the moon in July of 1969. Von Braun had joined Fairchild Industries as vice president for engineering and development after he retired from NASA. The struggle against Soviet Communism was still the guiding principle behind American foreign policy, which meant that while the public knew that von Braun had been a Nazi scientist, people didn't talk about it much.

Von Braun's crowning achievement was the landing of Apollo 11 on the surface of the moon on July 20, 1969, with astronauts Neil Armstrong and Buzz Aldrin.

But after the American landing on the moon, given the problems of Vietnam and turmoil at home, public interest in space flight dimmed. Because of the ensuing budgetary constraints, von Braun retired from

NASA on May 26, 1972, and went to work for Fairchild Industries, where he met Carol Rosin.

▲   ▲   ▲

Shortly after going to work for Fairchild in 1973, von Braun learned he had kidney cancer, which was not treatable at that time. He had only a few years to live.

Rosin says that in her very first meeting with von Braun, as tubes drained fluid out of one side of his body, he told her, "You will be responsible for keeping weapons out of space. I'm dying of cancer and you have to take over for me."[174]

Over the next few years, von Braun explained to Rosin the plan of the American military-industrial complex, and why it had to be opposed. Perhaps the experience of watching Hitler rise to power in Germany, and von Braun's own approaching death, made him want to prevent a similar madness in America.

He told Rosin that the United States would make Communists the first threat. To heighten their fear, Americans were told that they already had killer satellites in orbit, and we were behind them. Von Braun told her that was nonsense. The Communists were an exaggerated threat, and the failure of that system would soon be clear for the world to see.[175]

After the Communists, the threat would be terrorism. After the terrorists, the next threat would be asteroids.[176] Space-based weapons would be justified to protect us against threats like the meteor that killed the dinosaurs. The truth is, however, such defenses could be built for a relatively low cost and easily deployed. A couple of nuclear bombs in the path of any large body would be more than enough to deflect it from a collision with Earth. A sensible defense costing two or so billion dollars could do the job, although the chances of such an event's happening in the next few centuries is pretty close to zero.

Still, it's better to be safe than sorry.

Then von Braun told Rosin that the final threat would be the aliens. Over and over again, he told her, "Remember Carol, the last card is the alien card. We're going to have to build space-based weapons against aliens. And all of it is a lie. A lie."[177]

Michael and I discussed Rosin's account with her, but she did not want it recorded. Instead, she directed us back to her public testimony and said all of it was still accurate.

It's important to note that Rosin claims that von Braun never went into any detail about whether these aliens for which a space-based weapons platform needed to be developed were real. However, she claims that the look in his eyes as he talked about the phony alien threat hinted that there was a big secret that he could not divulge. All he would talk about was the list of "enemies" that the military-industrial complex was compiling, and how they'd be rolled out in the decades to follow. He didn't mention a specific timeline in which these events would take place, but claimed it would be faster than anybody could possibly imagine.

In the last years of his life, von Braun repeatedly mentioned to Rosin that there would be an education strategy to terrify the public and decision makers into green-lighting a space-based weapons system.

Von Braun wanted Rosin to be the counter to this deception.

She would educate the public and these same decision makers about the scare tactics being deployed against them, and that any such system would be too costly, unworkable, and ultimately destabilizing, which also would serve the interests of the military-industrial complex.

A few months after she'd first met with von Braun, Rosin was asked to give a speech for him in July of 1974 at Chicago's McCormick Hall before eighteen thousand people. The speech would be broadcast all over the world, as it would coincide with a live satellite demonstration.

Von Braun gave her five books as well as some pink index cards with notes. Rosin expressed her nervousness to him, especially about plowing through the five books he'd given her. "Don't worry," he said. "You don't really have to read all of these books, because I'm going to transmit the speech to you."

"How are you going to transmit the speech to me?" Rosin asked. "You're going into the hospital."

"I'm going to transmit it as though I'm talking to you on the telephone," he said.[178]

Rosin thought he must be joking, and with her five books and pink index cards with notes, made her way to Chicago. She memorized the notes as best she could, but when she started to speak, she realized that she was perspiring so heavily that she couldn't make out what was written on them.

At that moment, she began to hear von Braun's voice in her left ear, calmly telling her what to say, even though she didn't have an earpiece. She found herself talking about the Ku-band and the C-band, satellite-frequency concepts that she didn't understand but was explaining because von Braun was telling her what to say.

Later she would learn that this technology had a name, Voice of God, and that it had been used in many nefarious ways, especially among those with mental health problems, to do the bidding of the deep state.

When Rosin saw von Braun later and asked how he'd accomplished the feat, he just laughed. It was at this moment that Rosin realized there was an entire world of which she was unaware, run by people with access to futuristic technology that they did not share with the public. It was as if she'd been at the kids' table at Thanksgiving for her entire life and had now been allowed to hear what the grown-ups were talking about at the adults' table. Rosin kept this story private for many years, not sharing it until the early 2000s, because she knew how crazy it would sound to most people.

Rosin recalls that in 1977, she was in a meeting at Fairchild Industries, and there were charts on the walls with lots of names, including Saddam Hussein and Muammar Gaddafi, and maps of the Middle East. The people in the meeting were talking about the terrorists they expected to be fighting in the next few decades, and there was general agreement that there would someday be a war in the Persian Gulf involving the United States. It was assumed that this would be a time after the Soviets were no longer a concern, and America would need space-based weapons to defend against these future terrorists.[179]

Rosin stood up in the meeting and said, "I'd like to know why we're talking about these space-based weapons against these enemies. I'd like to know more about this. Would somebody tell me what this about?"

Nobody answered her, so she continued. "If nobody can tell me why we're planning a war in the gulf, when there's a certain amount of money in a budget so that you can create the next set of weapons systems, that will be the beginning of the sell to the public about why we need space-based weapons." She looked around the room, but nobody responded. Everyone was too busy delightedly planning the next conflict. After a moment, she said, "Consider this my resignation," and walked out.[180]

The meeting was filled with what Rosin refers to as the "revolving door people"—those who had once worn military uniforms but now were dressed in gray pinstripe suits and red ties as the highly paid assets of military contractors, or of some other government agency and sometimes even a university. It startled Rosin to realize the amount of influence the military-industrial complex wielded over the levers of power. She is adamant in her claim that military contractors for Fairchild Industries and others planned the first Gulf War, which kicked off in 1990, in 1977.[181]

▲   ▲   ▲

The 1980s were both a terrifying and an extremely exciting time. The election of Ronald Reagan to the presidency in 1980, with his strong anti-Communist rhetoric, made many people view a nuclear exchange with the Soviets as a much more likely scenario than previously. However, unlike many conservative cold warriors, Reagan appeared genuinely concerned about the policy of mutually assured destruction between the superpowers; he believed it was not a way for humanity to move forward.

In 1977, after von Braun died, Rosin founded the Institute for Security and Cooperation in Outer Space, a Washington, DC–based think tank dedicated to carrying on von Braun's dream of a peaceful future in space. Since that time, she has debated generals and congressional representatives, and testified before Congress and the Senate. Despite having met with people from over a hundred countries, she doesn't feel she's been able to identify those people who are pushing this space-based weapons systems.[182]

They remain in the shadows.

However, Rosin has concluded that these efforts are based on a few people's making a lot of money and gaining power. During the Reagan administration, this concept was sold to Americans as the Strategic Defense Initiative, dubbed "Star Wars" by the media, with the stated intention of being able to defend against a Soviet missile strike and undercut the policy of mutually assured destruction between the two superpowers. However, late in Reagan's second term, he went before the United Nations and asked the world to imagine how the squabbling nations of the world would quickly drop their disputes in the event of an alien invasion. Here is the relevant portion of that curious speech:

> I have spoken today of a vision and the obstacles to its realization. More than a century ago, a young Frenchman, Alexis de Tocqueville, visited America. After that visit he predicted that the two great powers of the future would be, on one hand, the United States, which would be built, as he said, "by the plowshare," and on the other, Russia, which would go forward, as he said, "by the sword." Yet need it be? Cannot swords be turned to plowshares? Can we and all nations not live in peace? In our obsession with antagonisms of the moment, we often forget how much unites all the members of humanity. Perhaps we need some outside, universal threat to make us recognize this common bond. I occasionally think how quickly our differences worldwide would vanish if we were facing an alien threat from outside this world. And yet, I ask you, is not an alien force already among us? What could be more alien to the universal aspirations of our peoples than war and the threat of war?[183]

Was this simply an instance of Reagan's optimistic rhetoric, the way the former actor could often see around corners, using vivid images to convince the public that a better path might lie ahead? Or was this a peek behind the curtains at—in essence, a wink to—von Braun's claim that two alien threats (asteroids and aliens) would be presented to the public?

It was in the 1980s, in the shadow of Reagan's claim that the Soviet Union was the "focus of evil" in the modern world, as well as hoping for a way out of the insane policy of mutually assured destruction, that Rosin made her greatest gamble. As an article in the December 2021 issue of the *Bulletin of the Atomic Scientists* details the effort:

> The plan was simple: Give the psychedelic drug MMDA (popularly known as Ecstasy) to Soviet scientists and military personnel set to negotiate with US President Ronald Reagan in 1985, thereby injecting empathy and cross-cultural understanding into the nuclear peace process.

So, that's just what Rick Doblin and Carol Rosin said they did— and they still believe introducing psychedelics into nuclear negotiations can produce positive results.[184]

One might be inclined to dismiss the story Rosin tells, except for the abundant documentation of the claim as well as the statements of some of the leading figures of the time.

In 1986, the year after Doblin provided Roslin with 1,000 doses of MDMA for her trip to Moscow, he founded the Multidisciplinary Association for Psychedelic Studies (MAPS)— a nonprofit that has since become a powerhouse behind pioneering studies on the use of psychedelics for the treatment of PTSD, alcoholism, and mental health issues.[185]

The claim is that psychedelics open up individuals to empathy and compassion for those they may have viewed in a negative light. What greater need for such a tool would there be than in the field of international affairs, which impact the lives of millions of people? Rosin certainly does have a knack for finding herself at some of the more consequential hinge-points of history.

Rosin feels we have come to the final phase of the plan that von Braun warned her about from his hospital bed in 1974, the alien threat.

She undertook an extensive investigation of the UFO phenomena, looking at thousands of years of possible visitations and recorded history— including more recent revelations of honest military and industry people with UFO experiences, whether those were sightings or work on crashed alien vehicles—and she came to some firm conclusions.

In her opinion, the aliens could have chosen to destroy humanity at any moment in the past several thousand years but have refrained from any hostile actions. Nothing she has seen suggests that aliens pose the slightest threat to the human race.

Rosin believes that the alien technology being hidden from mankind has the potential to create a world of prosperity and peace, as well as to give humanity the tools to clean the planet of pollution and to create abundant free energy, allowing for the development of a more peaceful civilization that could make its way to the stars. She believes that the alien technology, if opened to the public, would allow quantum leaps in transportation, communication, healing, and life extension.

In her opinion, humanity has reached a pivotal point in our evolution. Will we as a species allow ourselves to remain hostage to the warmongers and secret keepers? Or will we break free from their grasp and make the leap to become a peaceful civilization?

The dream of peace in space seemed so close, despite the COVID crisis and the uncertainty of the 2020 presidential election.

On February 4, 2021, TASS (the Russian news agency) and Rosin's website ran the following article:

> The Commander of the U.S/ Space Force, General John Raymond, advocates a professional dialogue with Russia and China on military activities in near-Earth space, as he reported while answering questions from a TASS reporter on Wednesday at an online briefing for journalists....

> According to him, by Spring, the armed forces entrusted to him will number 6,400 military personnel and about 10,000 civilians. It is assumed that the Space Forces will consist of approximately 16,000 military personnel from the Air Force and Navy...

> General Raymond says, "There's been a long history, even in the height of the Cold War, there was great collaboration.... Whenever countries disagree, there's always been agreement on partnership on the civil use of space. NASA and the Space Force can work together to create new norms in space."[186]

However, on February 24, 2022, Russia invaded Ukraine, touching off a brutal war that continues to this day. There has thus far been no significant movement between the United States and Russia on keeping space from militarization.

⋏ ⋏ ⋏

The Russian invasion of Ukraine has made Rosin increasingly pessimistic about whether humanity will make the right choice. In her unrecorded interview with us, she declared she is retiring from her public activism and moving with her husband to Oregon, and has made sure she has stockpiled supplies for whatever terror those who lurk in the shadows are preparing to unleash against us.

"Maybe it will take somebody from outside the UFO community, somebody who isn't associated with the feuds of ufology, and certainly isn't part of the deep state, to tell the real story," she told us in that unrecorded interview.

CHAPTER SIX

THE GREER ARCHIVE

O ne of the courtesies Steven Greer extended to us was to have his office send me a flash drive with what he considered to be the most persuasive evidence of the UFO phenomenon, as well as of our government's complicity in covering up the information.

Whether one agrees with Greer's point of view or not, we would argue that he is perhaps the most consequential voice in the field today, and it's important to understand his point of view as to the creation of the UFO security state.

The first document that caught our attention was a purported briefing document for then president-elect Dwight D. Eisenhower, which was dated November 18, 1952. Because of the importance of this memo in understanding what may have developed over the past seven decades, the text portion of the document is provided below in its entirety.

> OPERATION MAJESTIC-12 is a TOP SECRET Research and Development/Intelligence operation responsible directly and only to the President of the United States. Operations of the project are carried out under control of the Majestic-12 (Majic-12) Group which was established by special classified executive order of President Truman on 24 September , 1947, upon recommendation by Dr. Vannevar Bush and Secretary James Forrestal. (See Attachment "A".) Members of the Majestic-12 Group were designated as follows:
>
> Adm. Roscoe H. Hillenkoetter
> Dr. Vannevar Bush
> Secy. James V. Forrestal*
> Gen. Nathan F. Twining
> Gen. Hoyt S. Vandenberg

Dr. Detlev Bronk
Mr. Sideny W. Souers
Mr. Gordon Gray
Gen. Robert M. Montague
Dr. Lloyd V. Berkner

The death of Secretary Forrestal on 22 May, 1949, [a suspicious "suicide," as he supposedly fell to his death from the window of a mental hospital to which he had been confined for a nervous breakdown] created a vacancy which remains unfulfilled until August, 1950, upon which date Gen. Walter B. Smith was designated as a permanent replacement.

On 24 June, 1947, a civilian pilot flying over the Cascade Mountains in the State of Washington observed nine flying disc-shaped aircraft traveling in formation at a high rate of speed. Although this was not the first known sighting of such objects, it was the first to gain widespread attention in the public media. Hundreds of reports of sightings of similar objects followed. Many of these came from highly credible military and civilian sources. These reports resulted in independent efforts by several different elements of the military to ascertain the nature and purpose of these objects in the interests of national defense. A number of witnesses were interviewed and there were several unsuccessful attempts to utilize aircraft in efforts to pursue reported discs in flight. Public reaction bordered on near hysteria at times.

In spite of these efforts, little of substance was learned about the objects until a local rancher reported that one had crashed in a remote region of New Mexico located approximately seventy-five miles northwest of Roswell Army Air Base (now Walker Field).

On 07 July, 1947, a secret operation was begun to assure recovery of the wreckage of this object for scientific study. During the course of this operation, aerial reconnaissance discovered that four small human-like beings had apparently ejected from the craft at some point before it exploded. These had fallen to earth about two miles east of the wreckage site. All four were dead and badly decomposed due to action by predators and exposure to the elements during the approximately one-week time period which had elapsed before their discovery. A special scientific team took charge of removing these bodies for study. (See Attachment "C".) The wreckage of the craft was also removed to several different locations. (See Attachment "B".) Civilian and military witnesses in the area were debriefed, and news reporters were given the effective cover story that the object had been a misguided weather research balloon.

A covert analytical effort organized by Gen. Twining and Dr. Bush acting on the direct orders of the President, resulted in a preliminary consensus (19 September, 1947) that the disc was most likely a short range reconnaissance craft. This conclusion was based for the most part on the craft's size and the apparent lack of any identifiable provisioning. (See Attachment "D"." A similar analysis of the four dead occupants was arranged by Dr. Bronk. It was the tentative conclusion of the group (30 November, 1947) that although these creatures are human-like in appearance, the biological and evolutionary processes responsible for their development has apparently been quite different from those observed or postulated in homo-sapiens. Dr. Bronk's team has suggested the term "Extra-terrestrial biological entities", or "EBEs", be adopted as the standard term of reference for these creatures until such time as a more definitive designation can be agreed upon.

Since it is virtually certain that these craft do not originate in any country on Earth, considerable speculation has centered around what their point of origin might be and how they got here. Mars was and remains a possibility, although some scientists, most notably Dr. Menzel, consider it more likely that we are dealing with beings from another solar system entirely.

Numerous examples of what appear to be a form of writing were found in the wreckage. Efforts to decipher these have remained largely unsuccessful. (See Attachment "E".) Equally unsuccessful have been efforts to determine the method of propulsion or the nature or method of transmission of the power source involved. Research along these lines has been complicated by the complete absence of identifiable wings, propellors, jets, or other conventional methods of propulsion and guidance, as well as a total lack of metallic wiring, vacuum tubes, or similar recognizable electronic components. (See Attachment "F".) It is assumed that the propulsion unit was completely destroyed by the explosion which destroyed the craft.

A need for as much additional information as possible about these craft, their performance characteristics and their purpose led to the undertaking known as U.S. Air Force Project SIGN in December, 1947. In order to preserve security, liaison between SIGN and Majestic-12 was limited to two individuals within the Intelligence Division of Air Material Command whose role was to pass along certain types of information through channels. The operation is currently being conducted under the name BLUE BOOK, with liaison being maintained through the Air Force officer who is head of the project.

On 06 December, 1950, a second object, probably of similar origin, impacted the earth at high speed in the El Indio-Guerrero area of the Texas-Mexican border after following a long trajectory through the atmosphere. By the time a search team arrived, what remained of the object had been almost completely incinerated. Such material as could be recovered was transported to the A.E.C. Facility at Sandia, New Mexico, for study.

Implications for the National Security are of continuing importance in that the motives and ultimate intentions of these visitors remain completely unknown. In addition, a significant upsurge in the surveillance activity of these craft beginning in May and continuing through the autumn of this year [which takes into account the Washington, DC rash of sightings] has caused considerable concern that new developments may be imminent. It is for these reasons as well as the obvious international and technological considerations and the ultimate need to avoid a public panic at all costs, that the Majestic-12 Group remains of the unanimous opinion that imposition of the strictest security precautions should continue without interruption into the new administration. At the same time, contingency plan MJ-10949-O4P/78 (Top Secret—Eyes Only) should be held in continued readiness should the need to make a public announcement present itself. (See Attachment "G".)[187]

The question any researcher or reader would have after going through this document is whether it's authentic. No sources we consider credible have answered this question.

If it's a forgery, it's definitely of high quality. Greer and Michael Mazzola consider it authentic.

Considering the document from the perspective of a writer, the question is whether it has an internal consistency and can be logically connected to other credible sources. Based on books and publications like *Life* magazine, it's clear that beginning in 1947, something different (real or imagined) was being observed in the skies over the United States. And when one considers the year 1952, with the Air Force in the spring suggesting a greater openness to the phenomenon and then seeming to clamp down after the summer sightings over Washington, DC, this document, allegedly composed in November of 1952 to brief Eisenhower, seems consistent with how a government bureaucracy would respond to such a mysterious threat.

One might say the overall theme of the report to Eisenhower was "we don't know what's happening." The creatures were human-like in appearance but didn't seem to have evolved through the normal evolutionary processes. We're speculating at this point, but we imagine the military experts would have considered that any apex species on a planet would have achieved the role by means of a robust physical stature that made them capable of fending off all but the largest predators. The average human male weighs somewhere between 150 and 200 pounds, making him larger than most land mammals. Because of our social nature, a group of forty to sixty humans, with perhaps ten to fifteen well-armed warriors, is enough to fend off any of the large predators, such as lions, tigers, and bears.

The military men, looking at these creatures as potential warriors, would likely have found them to be deeply unimpressive.

What was impressive was the craft in which they flew, although it didn't seem to be a long-range craft, as they had no provisions. As to the power source, the military men were unsuccessful in determining anything useful. They were also similarly unsuccessful in deciphering what seemed to be writing. Given all of these considerations, the Majestic-12 group was unanimous in the belief that this information had to be held secret to prevent a panic.

An additional Majestic-12 document (undated), reveals the projects Majestic-12 undertook in the years that followed:

PROJECT AQUARIUS

(TS/ORCOM), (Proword): Grudge contains sixteen (16) volumes of documented information collected from the beginning of the United States Investigation on Unidentified Flying Objects, (UFOs) and Identified Alien Crafts, (IACs). The project was originally established in 1953 by order of President Eisenhower, under control of the CIA and MAJI. In 1960 the project's name was changed from project SIGN to project AQUARIUS. The project was funded by CIA funds (non-appropriated). The project assumed full responsibility for investigation and intelligence of UFOs and/or IACs, after December 1969 when Project Grudge and Blue Book were closed. The purpose of Project Aquarius was to collect all scientific, technological, medical and intelligence information form UFO and IAC sightings and contacts with Alien Life Forms. These orderly files of collected information have been used to advance the United States Air Force Space Program (Not NASA).

Aquarius is a project which compiled the history of Alien presence and their interaction with HOMO SAPIENS upon the planet for last 25,000 years and culminating with the BASQUE PEOPLE (PAIS BASCO) who live in the mountainous country on the border of France and Spain and the Assyrians (or Syrians, originally from the Syrius Star.)

(TS/ORCOM)

The preceding briefing is a historical account of the United States Government's investigation of Aerial Phenomena, recovered Alien Craft, and contacts with extra-terrestrial life forms.

THE PROJECTS UNDER "PROJECT AQUARIUS"

1. (TS/ORCOM) PROJECT PLATO: (Proword: Aquarius) Originally established as a part of Project SIGN in 1954. Its mission was to establish Diplomatic Relations with Aliens. This project was successful when mutually acceptable terms were agreed upon. These terms involved with exchange of technology for secrecy of Alien presence and noninterference in Alien affairs. Aliens agreed to supply MAJI with a list of names on a periodic basis. This project is continuing at a site in New Mexico.

2. (TS/ORCOM) PROJECT SIGMA: (Proword: Aquarius) Originally established as part of Project SIGN in 1954. Became a separate project in 1976. Its mission was to establish communication with the Aliens. This project met with positive success (SIC). In 1959, the United States established primitive communications with Aliens. On April 25, 1964 a USAF Intelligence Officer met with Aliens at Holloman Air Force Base, New Mexico. The contact lasted for approximately three hours, after several attempted methods of communications the Intelligence officer managed to exchange basic information with the Aliens. This project is continuing at a site in New Mexico.

3. (TS/ORCOM) PROJECT REDLIGHT (Proword: Grudge) Originally established in 1954. Its mission was to test and fly a recovered Alien Aircraft. First attempts resulted in the destruction of the craft and the death of the pilot. This project was resumed in 1972. The project is continuing in Nevada.

4. (TS/ORCOM) PROJECT SNOWBIRD (Proword: Redlight) Originally established in 1954. Its mission was to develop, using conventional technology and fly a "Flying Saucer" type craft for the public. This project was successful when a craft was built and flown in front of the PRESS. This project was used to explain UFO sightings and to divert the public's attention from Project Redlight.[188]

Again, it's difficult to assess the credibility of such a document. For the moment, let's withhold judgment on Project Plato (diplomatic relations with aliens), on Project Sigma (communication with aliens), and on Project Redlight (flying one of their crashed aircraft).

But did something like Project Snowbird (building a "flying saucer" using conventional technology to show to the press as an explanation for some UFO sightings) ever exist?

It did, and one needs only to go to the website for the Army Transportation Corps to get more information about the program. The craft was called the Avrocar, and here's what you'll find on the website:

> Developed in the 1950s by the Canadian firm AVRO Aircraft Limited, the Avrocar was meant to take-off and land vertically, cruise at low altitudes on a cushion of air, and accelerate to high speeds at higher altitudes.
>
> Two prototypes were built and resembled flying saucer UFOs. The Avrocar could reach 35 mph in test flights, but the aircraft became unstable at heights greater than three feet. The military scrapped the program in 1961.[189]

What is a regular consumer of news supposed to make of this information? A disc is about the stupidest design possible for a conventionally powered craft. The only way a disc would make sense is if there were a power source that could compensate for the lack of stability in the air when traveling at high speed.

If one were to look at the first two alleged programs, Project Plato (establishing diplomatic relations with aliens) and Project Sigma (establishing regular communications with aliens), the US government's response to this unique situation would seem reasonable.

In fact, if the government hadn't started figuring out how to deal with this potential threat by gathering more information, one would question their competence.

But these documents raise another troubling question: If these aliens are so advanced and have been visiting us for so long, why do they have a curious inability to communicate effectively with government officials? Many UFO researchers claim that the difficulty in communication is due to the fact that these aliens are so much more advanced than we are that it's equivalent to a human trying to communicate with an ant. However, contactees/abductees such as Betty and Barney Hill reported that the aliens had little difficulty communicating with them telepathically.

It strikes us as wishful thinking to ascribe all difficulties in the alien-human relationship to humanity, and all the good things to the aliens. We imagine that the Aztec ruler Montezuma made the same mistake when his series of missteps allowed Hernán Cortés and his relatively small band of mercenaries and native allies to overthrow the Aztec empire.

One of the most well-respected researchers in the UFO field is Timothy Good, and in his book *Earth: An Alien Enterprise*, he reviews the evidence for an early effort by the United States government, and particularly by General (and later President) Dwight D. Eisenhower, who was at the very center of the government's response to the UFO issue.

> In Mexico in 1970, the publisher Guillermo Mendizabal Elizalde told Ribera [a respected Spanish UFO researcher] that while attending a gathering during which a title was presented to Miguel Aleman Valdes, President of Mexico (1946-1952), the subject of flying saucers came up. Aleman listened in silence. But on being asked for his opinion, he revealed that when General Eisenhower had visited Mexico shortly before becoming president, he told Aleman that he had once been taken to an air base in the Southwest United States, where he had been shown "a flying disc and the cadavers of several of its crew members."

> This report was confirmed by Leonard Stringfield, a former U.S. Air Force intelligence officer, who learned from Dr. Robert Carr, a University of South Florida professor, that in 1948, General Eisenhower had been taken to see a craft and bodies captured at Aztec, New Mexico in March 1948, and it was on his command that the secrecy lid was clamped down on the subject, and rigidly enforced. Since another crashed disc had reportedly been recovered that same year just across the Mexican border from Laredo, Texas, it is likely, in my view, that this might have prompted Eisenhower's revelation to Aleman.[190]

If Good's account is to be believed, Eisenhower also likely would have known of the government's secret program to better understand the capabilities and intentions of these visitors.

In the chapter on Eisenhower's involvement with the extraterrestrial issue, Good points to a curious few days in the Eisenhower presidency, from February 9 to February 13, 1955, when Eisenhower allegedly went quail hunting for a few days at Milestone Plantation in Georgia, during which time he supposedly caught a cold and decided to convalesce for a few days. Many members of the press accompanied the presidential entourage on the trip down to Milestone Plantation for the alleged quail hunt, but they were not permitted on the grounds.

> With its two thousand acres of prime bird-hunting land, Milestone Plantation suited Eisenhower's requirements for privacy. With the exception of a previously arranged photo op, none of the media was allowed on the grounds. "The plantation was so secure," Art told me, "that Ike was able to go there four times in the 1950s. This was his second trip. The press was housed eight miles away at Scott Hotel in Thomasville. James Hagerty, Ike's press secretary, gave daily international news and briefings at the hotel lounge...."

> The hunting party reached the lodgings at Milestone, changed quickly and reached the hunting area at about 17:30. Media attendees included well-known journalist Ed Darby of *Time*, John Edwards of ABC, and William Lawrence of *The New York Times*. Why all the prestigious press for a quail shoot? Perhaps the reason lay in the fact that a week earlier, Josef Stalin's replacement leader, Georgi Malenkov, had been forced to resign and was replaced by Marshal Nikolai Bulganin.[191]

In his book, Good comments that given such international tensions, it seemed an unusual time for Eisenhower to go quail hunting. Also, the large press pool was suspect, and Eisenhower's subsequent alleged cold provided the perfect cover story for him to meet with alien representatives in New Mexico.

> On the day following arrival at Thomasville, James Hagerty reported that Eisenhower had come down with "the sniffles"

and would be staying in for a while. He was not seen again for some thirty-six hours, having secretly been taken to Spence Field and flown in Air Force One to Holloman Air Force Base, New Mexico, together with a team of Secret Service aides and supervisors.[192]

As Good says, we have what appears to be a perfect cover story, allowing Eisenhower to slip away from public view to deal with this issue. Good believes that's exactly what Eisenhower did, and quotes from several alleged eyewitnesses, starting with Airman Second Class Manuel W. Kirklin, who was having lunch with his wife at the hospital commissary, located on the base. About thirty minutes later, another man at the commissary told Kirklin about a flying disc that had landed at the base. Kirklin's wife suggested he go out to the front of the hospital to take a look.

Concerned at leaving his post without permission, Kirklin asked the head nurse if he could go. She consulted the doctor, but permission was refused. Later, the airman happened to be walking behind two pilots and overheard their conversation regarding the event. One of the two men, an Officer of the Day, was responding to questions from the other pilot relating to Eisenhower's visit. Kirklin asserts the officer explained that after Air Force One landed, it had turned around and remained on the active runway. The base radar had then been turned off, after which two discs had approached the base at low altitude via the White Sands National Monument.

"One hovered overhead like it was protecting the other one," explained the officer. "The other one landed on the active runway in front of [Eisenhower's] plane. He got out of his plane and went toward it. A door opened, and he went inside for forty or forty-five minutes." Asked by the other pilot whether he had seen the aliens, the officer replied that he had not: they stayed inside. Eisenhower then returned to his plane.[193]

This makes sense in the framework of increasing saucer sightings and crashes since the 1940s, as well as the growing panic of the 1950s, culminating with the sightings over Washington, DC, in the summer of 1952. If there was increased alien activity on Earth, both sides would have a vested interest in making sure things did not get out of hand.

Good includes accounts from service members who recall Eisenhower's visit to Holloman, but not much about the alleged alien encounter. However, Good does include this account by Bill Larson (a pseudonym), a civilian electrician at the base, who wrote the following in a letter to his daughter a few months before his death, about what he had seen while replacing wire on the flight line during the Eisenhower visit.

"Well, I had my climbers on and I started to unbuckle them and was waiting to give them to the first volunteer when someone said I should do it, as people were used to seeing me up the poles anyway. So I started up with my back to the sun—a safety measure—which also put my back to the runway where I thought his Connie (Constellation) was. As I started up, some of the guys reminded me not to gawk, and I heard them laugh. A few minutes later, I heard someone shouting, and some guys tarring the hangar roof nearby started to run, pointing out to the runway. Then I heard our truck start up and some of the crew jumped in, with one or two running after it, and they were pointing out to the flight line. And so I decided to turn around on the pole to see what the ruckus was about—and I could not believe what I saw.

"There was this pie-tin-like thing coming at me about 150 feet away. I thought it was remote-controlled or something, twenty-five to thirty feet across. And I started down the pole as fast as I could go. I was up about forty feet, and I threw my climbing rope out, gave it slack and only touched the spike on each side of the pole three or four times even before I got to the bottom. While I was running toward the big hangar, I looked back and it had stopped, and was just sitting there."[194]

Larson continues his account, recalling that his buddies had teased him about coming down "faster than a fireman."[195] Then Good pivots to Larson's daughter recounting what her father had told her about the incident.

"Dad said that once the people got over the initial shock, many just stood and watched. He said it was a beautiful sight. It had an occasional wobble. He recalled that later that day many neon lights needed replacing.... They all thought it was one of our secret aircraft that the president had come to see. Dad said he never considered it was anything but ours until years later when [the subject] got publicized more—in the 1960s or so. It was only then that he understood what was so secret...."[196]

After including this account, Good, being the honest researcher that he seems to be, tells the reader he can find no official evidence that Eisenhower made such a trip. However, he does include interviews with those who were in Georgia (where Eisenhower was reportedly staying), and mentions a big party on February 11, 1954, at a local country club with the lonely national reporters who'd been hanging around for a few days with nothing to do, and with a bevy of local, available women (rich divorcees) who'd been invited supposedly to distract the reporters from the absence of the president.[197]

Goode continues this section by quoting from a May 2010 speech by retiring New Hampshire state representative Henry W. McElroy, Jr., who claimed that early in his career he'd viewed a document regarding what Eisenhower knew about UFOs and extraterrestrials.

"The document I saw was an official brief to President Eisenhower. To the best of my memory, this brief was pervaded with a sense of hope, and it informed President Eisenhower of the continued presence of extraterrestrial beings here in the United States of America. The brief seemed to indicate that a meeting between the president and some of the visitors could be arranged as appropriate and if desired.

"The tone of the brief indicated to me that there was no need for concern, since these visitors were in no way causing any harm or had any intentions whatsoever in causing any disruption then, or in the future.[198]

Good continues to provide accounts from other researchers, supporting his claim that Eisenhower met with aliens.

Art Campbell describes his correspondence with a man who claimed to have had access to highly secret archives of the U.K. Security Service commonly known as MI5—who asserted that Eisenhower had had meetings with two or three separate groups of aliens during his presidency. "In the 1953-1955 timeline," wrote the source, "the ET visitors had landed at several places and asked for a meeting with the leader of the most powerful country on Earth. The top item on the meeting agendas was continued nuclear research and testing with more and more powerful weapons."

Regarding the nuclear agenda, Campbell's MI5 source alluded to a Russian nuclear weapon test in 1951 (September 24) that was almost double the size of the first weapon in 1949 (August 29), adding that the visitors showed particular concern about the first hydrogen bomb in 1952.[199]

This would be a continuing narrative motif in UFO lore, that the aliens were concerned about our development of nuclear weapons.

The question would be, why?

Is it because our explosion of nuclear devices was causing some sort of disturbance across the universe, as some contactees/abductees claim the aliens who spirited them away for a period of time told them?

Is it because they are concerned about our welfare, and worry about the devastation we might cause to our planet and ourselves?

Or is it because they have studied us and they're frightened of our potential to challenge them? We might be far behind them at the moment, but humans have a remarkable capacity to quickly adapt and innovate.

Perhaps the aliens are not so worried about us destroying ourselves.

Maybe they're more worried about us challenging them among the stars.

By contrast, Mazzola believes humanity is under some type of galactic quarantine until we're spiritually mature enough to possess their level of technology.

The answer to this question is critical to the future of humanity and maybe even, the fate of other civilizations.

▲ ▲ ▲

In the materials Greer sent to us, he included more than eighty-five statements by witnesses relating to their UFO experiences.

We have decided to share portions of five of these reports, as we believe they paint a picture of what may be happening behind the scenes as various factions of our government try to respond to the UFO issue.

The first is the text of a November 9, 1978, letter from Mercury Seven astronaut (the Mercury program was the first to put Americans into space in single astronaut capsules) Gordon Cooper to George Ashley Griffith, who at the time was the UN ambassador from the Caribbean nation of Grenada, and who had developed an interest in UFOs:

> Dear Ambassador Griffith:
>
> I wanted to convey to you my views on our extra-terrestrial visitors popularly referred to as "UFO's" and suggest what might be done to properly deal with them.
>
> I believe that these extra-terrestrial vehicles and their crews are visiting this planet from other planets, which obviously are a little more technically advanced than we have here on earth. I feel that we need to have a top level, coordinated program to scientifically collect and analyze data from all over the earth concerning any type of encounter, and to determine how best to interface with these visitors in a friendly fashion. We may first have to show them that we have learned to resolve our problems by peaceful means, rather than warfare, before we are accepted as fully qualified universal team members. This acceptance would have tremendous possibilities of advancing our world in all areas. Certainly then it would seem that the UN has a vested interest in handling this subject properly and expeditiously.
>
> I should point out that I am not an experienced UFO professional researcher. I have not yet had the privilege of flying a UFO, nor of meeting the crew of one. I do feel that I am somewhat qualified to discuss them since I have been into the fringes of the vast areas in which they operate. Also, I did have occasion in 1951 to have two days of observation of many flights of them, of different sizes, flying in fighter formation, generally from east to west over Europe. They were at a higher altitude than we could reach with our jet fighters at that time.

I would also like to point out that most astronauts are very reluctant to even discuss UFOs due to the great numbers of people who have indiscriminately sold fake stories and forged documents abusing their name and reputation without hesitation. Those few astronauts who have continued to have a participation in the UFO field have had to do so very cautiously. There are several of us who do believe in UFOs and who have had occasion to see a UFO from the ground, or from an airplane. There was only one occasion from space which may have been a UFO.

If the UN agrees to pursue this project, and to lend their credibility to it, perhaps many more well qualified people will agree to step forth and provide help and information.

I am looking forward to seeing you soon.

Sincerely,
L. Gordon Cooper
Col. USAF (Ret)
Astronaut[200]

As part of the final launch of the Mercury program, Cooper was the last man to travel alone in space. The craft was launched on May 16, 1963, and circled Earth twenty-two times in a mission that lasted thirty-four hours and nearly twenty minutes, the longest American spaceflight at that time. It's worth noting that as one of the most famous Americans of his time, Cooper did not need the controversy that would inevitably accompany his curiosity about the UFO phenomena.

But fifteen years after all Americans knew his name, here he was talking about UFOs.

The Greer archive contains several pages from *Leap of Faith: An Astronaut's Journey Into the Unknown*, a book by Gordon Cooper with Bruce Henderson about Cooper's life, published in 2000. We bought a copy of the book.

An early mention of UFOs in the Cooper book comes courtesy of the German rocket scientist who later got America to the moon, Wernher von Braun, whom Cooper worked closely with in the Mercury program. As Cooper recalls:

Wernher confided to me that the Germans were testing more than rockets at Peenemunde [a military research center]. "Some of the craft we were developing," he said, "were far ahead of anything the rest of the world had or knew about."

"You mean jets?" I asked, thinking of the Luftwaffe's Me262, the world's first jet fighter.

He smiled a scientist's knowing smile. "You could almost not refer to them as planes. We flew several craft that were totally different. Very advanced principles were involved."

According to Jack, [a German test pilot] who flew some of these advanced craft, they included saucer-shaped vehicles with double-intakes and counter rotating fans and disks with some advanced propulsion systems. Jack said they had flown successfully. None of these craft surfaced after the war.[201]

From Cooper's telling of the story, one gets the sense that among the aeronautical and rocket-design community, the UFO phenomenon was something of an open secret, discussed within the community but not with the wider world. He shares details about two other former Nazi German rocket scientists who worked in the US space program and believed in UFOs.

Early German rocket scientist Hermann Oberth also ended up in the United States. He was hired in the mid-1950s as a consultant for the U.S. Army Ballistic Missile Agency and later by NASA. Oberth, whom I met through Wernher, had a brilliant mind that accepted few limitations to man's reach. He had strong views about UFOs, having been hired by the West German government to head a commission for the study of UFOs following the war. In the commission's final report, Oberth contended that some of the unexplained objects were "propelled by distorting the gravitational field, converting gravity into usable energy." He believed that there was "no doubt" that some of the unexplained objects were "interplanetary craft of some sort that do not originate in our solar system."[202]

If what Cooper alleges is true, it seems the Nazis must have also obtained some downed UFO wreckage (perhaps material from a supposed

1933 Italian UFO crash, said to be given to Hitler by his good friend and fellow dictator Benito Mussolini), and they were experimenting with it. Some researchers have alleged that the "foo fighters" (small flying spheres, often crackling with energy) Allied air crews saw over Europe were in fact using German technology based on reverse engineered alien technology, meant to confuse the Allied pilots.

Cooper continues with his account of other German rocket scientists who believed in UFOs.

> Another former top German scientist who agreed with Oberth was Walter Riedel, once chief designer and research director at Peenemunde. Riedel, who also came to this country to work in the U.S. space program, kept records of saucer sightings around the world. Convinced that some of the sightings had an "out-of-world basis," he offered several strong arguments to support his belief: the skin friction of the craft at speeds and altitudes observed would melt any known metals or nonmetals available; the high acceleration at which they flew and maneuvered would be intolerable for the crew; the many instances in which they had done things that only a pilot could perform but that no human pilot could stand; the fact that in most of the sightings there was no visible jet flame or trail, suggesting "no power unit we know of." From my own experiences and those I'd heard from pilot friends, I was in complete agreement with these findings.[203]

Three giants of German and American rocketry, Wernher von Braun, Hermann Oberth, and Walter Riedel, were convinced that UFO reports were real, and that they used a technology far beyond anything humans had created. It's remarkable that the questions engineers raised in the 1940s would still be relevant more than seventy-five years later.

When we consider this next story by Cooper, we can't help but recall Pulitzer Prize finalist Annie Jacobsen's preposterous story in her book *Area 51*, that the Roswell crash was of a craft designed in the Soviet Union and piloted by children surgically altered by Nazi physician Josef Mengele.

> Wernher told of a firsthand UFO demonstration that had been given to him and other German rocket scientists and U.S. mil-

itary personnel at White Sands, New Mexico on July 10, 1949. While tracking the test launch of a V-2 rocket at two thousand feet per second, scientists suddenly saw two small circular UFOs pacing the missile. One was seen to pass through the missile's exhaust and rejoin the other object. Then they quickly accelerated upward, leaving the missile behind. Seeing their capability for hovering around a speeding rocket and accelerating away from it with apparent ease left a vivid impression on all who witnessed the event.[204]

These aren't ghost stories from teenagers spending the night in some old haunted house, hyped up on fear and hormones, but the experiences of scientists and soldiers in the light of day testing what were the most advanced terror weapons of their time.

In chapter twelve of his *Leap of Faith* book, titled "UFOs at the United Nations," Cooper tells the story of two presidents and a presidential contender who saw UFOs. The first was Jimmy Carter.

Jimmy Carter filed a formal report while he was governor of Georgia describing his sighting of a UFO. He was leaving a Lion's Club in Leary, Georgia, when he noticed an object in the sky. A red-and-green "glowing orb" hurtled across the southwestern skies that evening in January 1969, vanishing a few minutes later. Carter told his story repeatedly in the years following, becoming the first major politician to risk being labeled a "UFO crackpot." Carter told the Southern Governors Conference a few years later: "I don't laugh at people anymore when they say they've seen UFOs. I've seen one myself."[205]

The second person who saw a UFO was Ronald Reagan, who defeated Carter in the 1980 presidential election. Reagan was a governor at the time of his sighting.

Ronald Reagan also had a UFO encounter, which I heard about from a good friend, Bill Paynter, a former air force pilot with about 45,000 hours of flying time in everything from fighters to bombers. Paynter provided the airplane that the state leased for Reagan when he was governor, and came along with it as chief pilot.

One evening, en route to Los Angeles, a UFO pulled up alongside the plane and sat off their wing. Governor Reagan and the rest of his party looked out their window at the object while Paynter did a number of cautious maneuvers to try to lose the saucer-shaped craft, which was uncomfortably close to the chief executive's plane. The UFO maneuvered with them for several minutes before darting out of sight.[206]

Probably the most entertaining account of a UFO encounter by a major politician is that of US senator Barry Goldwater, the 1964 Republican nominee for president.

U.S. Senator Barry Goldwater, a longtime pilot and air force reserve general, had his own encounter with a UFO one day over the Arizona desert. When he was governor, he used to fly an F-86, a single-seat fighter, on government and personal business—as he was governor he was answerable only to himself for the use of National Guard aircraft. During the space program, I had several informal meetings with Goldwater. At one of them, he told me about his UFO experience.

"I chased it all over the desert and couldn't get near it," he said. "Damnedest thing I ever saw. Made me a believer."

Goldwater described the UFO he played tag with that day as a "shiny saucer." It was capable of much greater speeds than his F-86, he said, and was able to stop on a dime and make turns that seemed to defy gravity.[207]

Further in the book, Cooper details his experience testifying before the United Nations about his UFO encounters. He told the UN about meeting luminaries such as J. Allen Hynek (originally the chief scientist on the government's Project Blue Book investigation into UFOs, who later became convinced there was a massive government cover-up), as well as other curious characters who claimed to have had direct telepathic contact with aliens.

As to the ultimate meaning of it all, Cooper offers no final answer, just a mind open to the possibilities.

▲ ▲ ▲

The accounts from Steven Greer's archive that I've chosen to highlight do not have Gordon Cooper's fame and visibility, but they tell remarkably similar stories of unusual encounters and official neglect. This is from Harry Jordan, and details what he saw as a radar operator in the fall of 1963 on board the USS *Franklin D. Roosevelt*:

> I was a radar operator trained at OCS [Officer Candidate School] in Newport, RI. I was trained in air search radar, and how to operate top secret equipment for electronic counter-measures that could listen in on Russian Soviet transmission and be able to detect the kind of aircraft or ship that certain radar emissions were coming from.
>
> In the fall of 1963, I was on mid-watch at about 2:30 am in the morning and we picked up this contact about 360 miles from the ship at about 80,000 feet. And it was moving very slowly, I'd never seen anything like it. At first I thought I was on surface search radar which is for ships, and instead it was on air search.
>
> Called the commander in and several other people were around. And my CO Commander Gibson at the time, said, "Jesus, Jordan! What is this?"
>
> We checked for IFF..."Friend or Foe" code which tells us if it's military or civilian. The UFO contact was out of the normal commercial flight pattern. The minute we turned on our active ECM...this thing started closing the ship at 4,000 miles per hour. In less than a minute and a half, it had closed the ship at least 180 miles. Then it stopped and hovered at about 60,000 feet and then it came down low and headed right for us.
>
> The captain turned the ship into the wind, launched two jet aircraft, the minute they turned on their conical scan radar, and this thing winked out. And when it winked out the jets flew around for about twenty minutes. Then we headed into the wind, and headed back onto our course, south by southwest of

the coast of Sardinia, and the straits of Bonifacio, just north of Sicily.

Within ten minutes, this thing came on the scope again, the signal was even stronger. I detected this thing at about twenty thousand feet. In those days, when we had the new Spa 50 repeater, the USS Franklin D. Roosevelt aircraft carrier and then there are other reports from other Navy personnel who were on board, who saw this thing happen.

After this happened, and I remember this very vividly, when I first saw the movie *Red October*, this Admiral comes in and shows his ID to the guy working on sonar and says "Now, this never happened."

Some guy came out of nowhere—I'd never seen him before— an intelligence officer who wasn't attached to our division. He looked at me and he said, "Jordan, this never happened!" I was told that I couldn't talk about it. I had a classified clearance.[208]

One gets the idea from this report that there are two classes of people in the government: the vast majority who are not let in on the secret, and a small number, like the intelligence officer who appeared out of nowhere, who are briefed on what is really going on.

▲   ▲   ▲

The next account comes not from the military but from the civilian aviation world and has echoes of the same intelligence involvement as the account of the radarman on the USS *Franklin D. Roosevelt*. It's an email to Greer.

Dr. Greer:

I retired from the FAA, Div. Mgr. Accidents, Evaluations and Incidents Div./FAA in 1988. A year before retiring I video-taped a radar play back of a UFO "chasing" a Japan 747 over Alaska for some 30+ minutes. Gave a briefing to the FAA administrator and then to the President's scientific staff, CIA and others.

They think they took all the material we had an advised "this never happened"…"we were never here".… I kept my originals and even have the 747's pilot report of the incident.

A friend of ours sent me a copy of your story in the local paper. That's how we came by your name. If you would like to see the video, want a copy or just talk about what happened, give us a call.…

John J. and Dori Callahan[209]

To recap, there was a Japan Airlines 747 traveling over Alaska, and it was shadowed for approximately a half hour, as recorded on radar. John Callahan, a witness, presented this information to the Federal Aviation Administration, the science staff of the president, the CIA, and other entities. These entities did not want him to talk about this information.

Time and again, we've gotten a picture of government agencies standing by to gather information about these incidents, as well as making efforts to keep the information from spreading too widely or, when the situation requires it, to discredit the witnesses.

The stories that follow go even deeper into this picture.

⋏　⋏　⋏

From May of 2000 comes a report that investigator Paul Nahay gave to Greer, about an individual named Harlan Bentley. This is how Nahay describes Bentley, with whom he had met several times:

He was a soldier at the Nike Ajax Missile Battalion. Mt. Zion, MD (near Olney, MD), and himself witnessed the crash of a saucer in May 1958 (easy month/year for me to remember, as I was born that month). While there he also tracked the secret SR-71 U.S. spy plane on radar, and was told by a general not to track it. He worked at Los Alamos. Lawrence-Livermore, Sandia Labs during '88-93.' He is an Electrical Engineer, specializing in radio frequencies, and has been on the road almost constantly, in demand by industry in the U.S. and abroad as a consultant on the Y2K thing, and now matters pertaining to radio frequencies. He almost had a PhD in Nuclear Engineering from Univ, of Maryland @ College Park (my own alma mater), was also pre-med, and studied also at George Washington Univ. He knows multiple languages....

He confirmed the existence of the "shadow government" (and had lots of bad things to say about them), and their back-engineering programs, as well as pointing out some specific things that were back engineered (he actually had in his possession a night-scope and an infrared scope, which I got to look through, and was astounded by!) He seems to think that something big will happen between now and 2012. He talked some about the nature and intentions of the different "races" of aliens, and how the U.S. military has worked with them. He has seen the black helicopters many times which have come out when he has had contact with aliens.[210]

This account highlights some of the things that are compelling about the quality of many witnesses in the Greer archive, such as the individual's years of government service, but also some troubling aspects, such as the claim of continued contact with aliens. In another portion of the report is an account of an "abduction" experience when Bentley was nineteen, followed by other "contacts," which had taken place with his agreement.

The average person might consider the possibility of individuals seeing unidentified objects in the sky, and yet the thought of continued contact might strike them as beyond the pale.

We do not know how to respond, except to say that among those who claim to have seen a UFO, this is a common pattern. These experiences often seem to generate something of a relationship between the abductee/contactee and the occupants of these crafts. In another part of the report, Nahay writes:

What he told me was fairly mind-blowing, but everything he said was related to something I already knew, simply confirmed what I knew. And, much of what he said absolutely confirmed what you yourself have been saying all along. (I was constantly telling him, "That's what Steven Greer says!") You'll be particularly gratified to know that he affirms that many abductions are in fact human-initiated, and also he described technically what you call "the crossing point of light" (he doesn't use that phrase of course, but seems to be describing exactly what you describe in your lectures and your book, the "frequency shifting" that allows inter-dimensional travel, etc.) He has heard of you, and thinks he's even seen the CSETI web site (which I showed him.)[211]

When investigating the UFO phenomena, one often gets the sense of being in a hall of fun house mirrors, with some providing accurate reflections and others providing distorted visions of reality.

Maybe we simply have to throw all the information down on the table and look at it, and only later determine what makes sense.

▲   ▲   ▲

The next was John W. Warner IV, the sixty-year-old son of longtime US senator John W. Warner of Virginia, who at one time was married to the actress Elizabeth Taylor.

In the witness statement he prepared for Greer on October 27, 2016, he comes off as charming and delightful, the type of person one might choose to sit next to at a dinner party. However, one might also question his grip on reality. Here's a sample of what he wrote:

> A pro racing driver for many years, I'm also a writer, humorist, filmmaker, and historian. And believe you me, I know our real history, not the fluffy, fake one in the "new and improved" fashion black box.
>
> How? Why?
>
> My heiress mother, an irascible sort and perennial thorn in my father's rump, is a free-thinking hippie, and has over the eons, balanced out my flag-waving, gun-totin' conservatism quite nicely indeed. She is a learned woman, pot brownie baker extraordinaire, never believed in the moon landing, among other inconvenient government half-truths. (Neither did Liz Taylor by the by. Very suspicious.) Forthwith, I will brief her on our genuine situation as best I can.[212]

Usually when analyzing the credibility of an individual, you simply investigate whether they are whom they say they are. There's no problem with that in this case, as Michael Mazzola has met with him and Greer has confirmed his background.

But just because the background checks out, does that mean they're telling the truth?

Warner continues:

> For the last 25 years I have known about the Majestic-12 project. When I showed the documents to my dad at the time, I demanded a veritable explanation. He denied them as a hoax vehemently, overly so.

> *Snap!* I knew then and there they were genuine, but I did nothing about it for eighteen years, my own Cold War naivete showing its flabby, pale backside. Those plucky Nellis Range boys must've known what they were doing, right?

> Since then I have independently confirmed everything in this silly, short-sighted "black empire" as we know it, through my Intelligence Community contacts and the handful of former military officers I trust. It all just fell into my lap like a fresh load of horse apples. And it snowballed. It's grotesque, abhorrent, absurd, and yes, *silly*.

> What a damn mess. Oh well. As my lovable racing buddy Paul Newman—another pie-sky, utopian Hollywood subversive like Liz—once said to me: "Them's the breaks, kid!"[213]

How can one not find that kind of description anything other than fascinating and compelling?

On the one hand, Warner is the son of a former United States senator. But on the other, he still can be as screwed up and delusional as any other human being. In his letter to Greer, he mentions that he has written a long memo covering everything he knows and is happy to send it to him. After a few more pages of his entertaining riff, Warner makes a very interesting claim:

> Included in my fancy pants expose is my grandfather Paul Mellon, OSS, who worked with Allen Dulles and Carl Jung

during WW2. Jung was OSS #488, and later briefed MJ-12 on metaphysics, the occult, and social psychology in order for them to better understand those pesky, skulking UFOs and their agenda-prone pilots on psychic autopilot. Most likely Jung's sage advice fell upon deaf ears and limp manhoods. Not much has changed I surmise.

Hmm, I like that. I'll put that in the Memo.

Anyway, dear old grandad was partially read in on the Nazi Bell program under SS General Hans Kammler, the extent of which I know not, but Paul was a sneaky sort. In the 1980's he told me after *several* stiff martinis that he and OSS London Station Chief David K.E. Bruce had seen a German antigravity flying saucer in the flesh with General Patton. At the time I thought he was just bragging and bullshitting me, and I forgot all about it. I thought he was referring to the unclassified jet engine ones that didn't really work, not the crackerjack torsion field Bell technology go-jobs we know of nowadays. Whoops. Gotta love those rotating thorium/mercury isotopes 'n' whatnot, right? Right.[214]

We did some research and can assert that Warner's grandfather was indeed the wealthy industrialist Paul Mellon, that Carl Jung did in fact consult for the Office of Strategic Services (OSS, the forerunner to the CIA) under OSS #488 in World War II (creating psychological profiles of Nazi leader Adolf Hitler and the state of the German mind), and that David K. E. Bruce was the London station chief for the OSS.

But none of that information is remotely helpful in figuring out whether Warner's claims are true. The best I can do is put the information in a mental file marked "Maybe True, Maybe False," and keep going through more information.

▲　▲　▲

This final account from the Greer archive isn't even about a UFO sighting. It was supposedly submitted by an Air Force veteran with a higher security clearance than top secret, who had been brought in to troubleshoot the photographic equipment of the Lunar Orbiter, an unmanned probe sent to the moon prior to the first Apollo landing to help pick a landing site. The anonymous writer sounds like somebody who worked in a technical capacity, but as with Warner's account, I can't come to a strong opinion about whether it's true or false.

An Airman escorted me into a darkroom. Inside, another young Airman assembled strips of high resolution 35 mm film into what is called a mosaic. He was placing side-by-side successively numbered photographic scans of the lunar surface, which had been transmitted back to Earth from the Lunar Orbiter. Each surface scan covered a narrow band of terrain, and successive orbits around the Moon were required to assemble a complete photographic image of the Lunar terrain.

The mosaic negative created by that process was then placed into a Resolution Enhancing Contact Printer. Photographic paper was placed on top of the negative, and an exposure begun. The negative was scanned by an electron light beam generated by a large Cathode Ray Tube, similar to the tube in a black and white Television set. The light beam was picked up by a photo-multiplier tube and, through a feedback loop, modulated by the various changes in density of the photographic negative, enhancing the contrast, brightness and resolution of the image in the process. The resulting 9.5 inch by18 inch high resolution contact print was then examined by a photographic interpreter or scientist, who viewed the images under a microscopic type viewer, analyzing the features and terrain of the Lunar landscape.

Left alone on the faint red light of the darkroom with the Airman and equipment, much of which I had never seen before, I began to question the technician, attempting to discern what the problem might be with the ailing contact printer. After a few minutes of investigation, it became clear that there was a problem with the electronic control circuitry, which was comprised of several small plug-in circuit modules. Having no spare parts on hand, it was clear that I would have to trouble-shoot each module on a component-by-component basis, a very tedious and time-consuming process at best. This was something that could not be done in the faint red light of a darkroom. The unit would have to be removed from the darkroom and taken into a more appropriate space to allow for the accomplishment of the task.

Talking with the airman on the other side of the room, questions floated in my head. I was curious and fascinated with the whole process. How were the signals from the Lunar Orbiter transmitted to the lab? How were they converted into images on photographic film? How were the images correlated and aligned into the final mosaic negative. I knew these were all questions that I should not ask, and yet at the same time, I was alone with an Airman who was obviously as enthusiastic as I was about his job. Under normal operating conditions, many other people would have been in the lab, part of the assembly line of production. But, here we were, all alone, so I began to ask all those questions.

After about thirty minutes of technical discussion and a complete rundown of the steps in the process, the Airman turned to me and said candidly "You know, they've discovered a base on the back side of the Moon!"

I said, "What do you mean?"

And again, he said, "They have discovered a base on the Moon!" And, surreptitiously, at the same time, dropped a photograph in front of me.

There it was, a mosaic print of the surface of the Moon, with some sort of geometric structures clearly visible. Scrutinizing the image, I could see spheres and towers. My first thought was, "Whose base is it? Then I realized the full implication: it was not anyone of this earth.

I did not dwell on the photograph—I quickly took it in visually and moved away in case someone else should enter the lab. I knew that I had been given a gift, information that I should not have seen. With my "position" being that of a dutiful Airman, I asked no further questions and went about my business, quietly thinking to myself that I couldn't wait to hear about this on the News in the next few days. I told myself, do whatever you can to get this thing fixed...so the world can see this and hear about it.[215]

It seems to me that either this account is from an individual doing his best to recall a long-ago incident, or it is a well-manufactured fake.

When one considers the archive of materials that Greer has generated over thirty years of investigating, a clear narrative comes into focus.

This narrative put forth by the Greer archive suggests that near the end of World War II, the government made efforts both to hide this information from the public and to develop effective countermeasures against the potential threat.

Under the umbrella of the Majestic-12 Project and working group, many programs were established, possibly culminating with President Eisenhower's meeting some aliens.

The information about UFOs was widely but privately shared within the space community, including with people like Nazi rocket scientist Wernher von Braun, Mercury astronaut Gordon Cooper, and fellow Mercury astronaut and later US senator John Glenn.

Jimmy Carter and Ronald Reagan saw UFOs, and Barry Goldwater chased one in his fighter jet. The son of a prominent United States senator supported the information and analysis that Greer brought forward, including one account that suggests we discovered an alien base on the far side of the moon in the late 1960s.

The mind boggles at all these possibilities. How can we ever be expected to figure out what is true?

In the next chapter, we will review certain allegations as detailed by a government report on historical UFO sightings released in March of 2024.

CHAPTER SEVEN

THE MARCH 2024 AARO REPORT ON UFOS

Volume I of the "Report on the Historical Record of US Government Involvement with Unidentified Anomalous Phenomena (UAP)" was finished in February of 2024 and cleared for publication the following month, on March 6.[216]

It's a curious and troubling document.

On the one hand, it seems intent on clearly dismissing the UFO phenomenon. On the other, it shows so many attempts by the government to do exactly that and yet apparently failing, that it strikes one as being like an alcoholic proclaiming right after waking up from his latest bender that he's definitely going to stop drinking.

The report lists twenty-five governmental investigations, starting with Project SAUCER in 1947 and ending with the All-domain Anomaly Resolution Office, established on July 15, 2022. That's a time frame of seventy-six years in which the public, and apparently a good portion of the military, as well as a few US presidents, have been chasing phantoms.

This is how the report opens:

> This report represented Volume I of the All-Domain Resolution Office's (AARO) Historical Record Report (HR2) which reviews the record of the United States Government (USG) pertaining to unidentified anomalous phenomena (UAP). In completing this report, AARO reviewed all official USG investigatory efforts since 1945, researched classified and unclassified archives, conducted approximately 30 interviews, and partnered with Intelligence Community (IC) and Department of Defense Officials responsible for controlled and special access program oversight, respectively.[217]

Let's talk about trying to solve all possible issues. "All Domain Resolution Office." That would involve the skies, oceans, lakes, rivers, streams, forests, mountains, cities, towns, hamlets, and maybe even underground caverns?

And what's going on with the government's having investigated this phenomenon since 1945? Weren't we supposed to believe that the modern UFO flap began with Kenneth Arnold's sighting in Washington state in June of 1947, followed by the Roswell crash in July of 1947? Something's unclear about the timeline here. Maybe this is support for the Trinity crash in August of 1945. The report continues:

> Since 1945, the USG has funded and supported UAP investigations with the goal of determining whether UAP represented a flight safety risk, technological leaps by competitor nations, or evidence of off-world technology under intelligent control. These investigations were managed and implemented by a range of experts, scientists, academics, military, and intelligence officials under differing leaders—all of whom had their own perspectives that led them to particular conclusions on the origins of UAP. However, they all had in common the belief that UAP represented an unknown and therefore, theoretically posed a potential threat of an indeterminate nature.[218]

All of these claims seem to be backed up by the historical record, especially with the military believing that these UFOs might pose some sort of threat to aviation and to humanity itself.

The executive summary, with boldface and italicized text as in the original, seems to slam the door on claims that UFOs represent some form of alien technology.

> *AARO found no evidence that any USG investigation, academic-sponsored research, or official review panel has confirmed that any sighting of a UAP represented extraterrestrial technology.* All investigative efforts, at all levels of classification, concluded that most sightings were ordinary objects and phenomena and the result of misidentification.... *AARO found no empirical evidence for claims that the USG and private companies have been reverse engineering extraterrestrial technology. AARO determined, based on all information provided to date, that claims involving specific people, known*

locations, technological tests, and documents allegedly involved in or
related to the reverse-engineering of extraterrestrial technology, are
inaccurate.[219]

Perhaps it's the lawyer in one of the coauthors, but we see some significant wiggle room in this document. At the same time, we acknowledge that a genuine scientific investigation will never conclude that something is untrue, simply that no evidence can be found for the claim. It is thus possible for a scientifically valid conclusion to also sound like evasiveness to a lawyer.

Let's examine the first claim, that no government investigation has ever found that a UFO sighting represented extraterrestrial technology. To us, that presented a statement of basic fact. How could anybody conclude, simply from a sighting, that an unidentified craft was definitely extraterrestrial?

All one could rationally conclude is that it had characteristics that do not fit any known aircraft. These craft don't exactly have "Alpha Centauri" painted on the side in order for anyone to better identify the star system from which they might originate.

We're a little confused by the statement that the investigation found "no empirical evidence for claims that the USG and private companies have been reverse engineering extraterrestrial technology." What's meant by "empirical evidence?"

As we understand the claims made by approximately thirty alleged whistleblowers, the government and private companies have been engaged in this secret program for decades.

Would these entities give up the information if simply questioned, the way one might ask a murder suspect, "Hey, did you kill this guy?" If the suspect replies, "No," that seldom means the investigation is over.

Usually, before a suspect is asked such a question, he or she is presented with a significant amount of evidence supporting the claim. It's unclear whether the investigators described in the report had the power to do anything more than ask questions.

As one reads further, the document becomes murkier.

AARO successfully located the USG and industry programs, officials, companies, executives, and documents identified by interviewees. In many cases, the interviewees named authentic

USG classified programs well-known and understood to those appropriately accessed to them in the Executive and Legislative Branch: however, the interviewee mistakenly associated these authenticated USG programs with alien and extraterrestrial activity.[220]

The whistleblowers correctly identified secret US programs, including those in the private sector, as well as the government officials and corporate executives involved in these programs, but were simply mistaken that they had anything to do with extraterrestrials.

If this were a murder case, that would be akin to saying the police were able to establish that the suspect was in the vicinity of the victim at the time of the murder, and witnesses saw the victim and suspect interacting, but those witnesses who claim they saw the suspect stab the victim were mistaken.

But AARO is saying, in effect, the witnesses are truthful except when it comes to the subject of extraterrestrials.

And the evidence of UAPs isn't limited to eyewitness testimony. There's sensor data as well, covered in a curious passage from the executive summary:

> Although many UAP reports remain unsolved or unidentified, AARO assesses that if more and better quality data were available, most of these cases also could be identified and resolved as ordinary objects or phenomena. Sensors and visual observations are imperfect; the vast majority of cases lack actionable data or the data available is limited or of poor quality.[221]

Wait a minute!

Is this report saying that many UFO reports remain unsolved? And that in addition to eyewitnesses' visual observations, there's sensor data supporting the observations? And the AARO's only response to that information is, "We think more data would allow us to solve these sightings as being of regular objects?"

The AARO group is making an enormous leap in logic: that if there were more information available (which could be said for just about any situation), they'd be able to prove there's nothing anomalous about the UAP phenomenon.

And in terms of burying the most explosive finding of the report, the AARO admits that there is sensor data that supports the visual sightings. How likely is it that the sensors would malfunction at exactly the same time as the witnesses were claiming to see something, and that all the sensors would malfunction in the same manner?

We'd say the probability is very low.

Moving on in the report, let's conduct a brief review of the twenty-five investigations the government has engaged in over the years to investigate UFOs.

The first program was Project SAUCER. The official date of its founding was December 30, 1947. From the report:

> Project SAUCER investigated one of the first well-known accounts provided by a private pilot, Kenneth Arnold. The pilot claimed that on June 23, 1947, while flying near Mount Ranier, Washington, he saw nine, large circular objects flying in a formation, objects that periodically flipped and were traveling at 1,700 miles per hour. He also compared the flight characteristics as the "tail of a Chinese kite." Arnold described their shape as "saucer-like aircraft." His account was picked up by several media outlets, and the term "flying saucer" emerged.[222]

The second investigation, Project Sign, began in January of 1948, and ended in February of 1949. Why did they need a second report, if the first one found nothing to worry about? From the results section of the report:

> The project evaluated 243 reported UFO sightings, and in February 1949, it concluded that "no definitive and conclusive evidence is yet available that would prove or disprove the existence of these unidentified objects as real aircraft of unknown and unconventional configuration." Project SIGN demonstrated that nearly all were caused by either misinterpretation of known objects, hysteria, hallucination, or hoax." It also recommended continued military intelligence control over the investigation of all sightings. It did not rule out the possibility of extraterrestrial phenomena.[223]

Project Sign evolved into Project Grudge, which ran from February to December of 1949. From the results section of the report:

Project GRUDGE investigated 244 reports of UFO sightings. It did not discover any evidence that the UAP sightings represented foreign technology; therefore, these findings did not pose a threat to U.S. national security. The report recommended that the organization be downsized and de-emphasized because it was believed Project GRUDGE's very existence fueled a "war hysteria" within the public. The USAF subsequently implemented a public affairs campaign designed to persuade the public that UFOs constituted nothing unusual or extraordinary. The stated goal of this effort was to alleviate public anxiety.[224]

A "public affairs campaign" with the specific aim of encouraging the public to buy into one side of the UFO issue?

The fourth investigation was Project Twinkle, which ran from the summer of 1949 to the summer of 1950. From the background section of the report:

Project TWINKLE was established in the summer of 1949 to investigate a series of UFO reports witnessed by numerous observers in Nevada and New Mexico. These UFOs were described as "green fireballs" streaking across the sky, moving in odd ways, and—in at least one account—the fireball navigated near an aircraft. The literature is not clear if Project TWINKLE was officially supported by the original Project GRUDGE, but it was managed by the USAF's Cambridge Research Laboratory. The goal of this investigation was to use multiple high-powered cameras near White Sands with the hope that if at least two images of the fireballs were captured, then their speed, altitude, and time could be observed.... The project was only able to secure one camera, which was frequently moved between locations following fireball reports, and no photographs of the fireballs were ever taken.[225]

The fifth investigation brought back Project Grudge, which had been dissolved in December of 1949. It existed from October of 1951 until March of 1952. From the results section of the report:

The new Project GRUDGE noticed that there was some cor-
relation between sightings and the publication of UFO stories
in the media. Capt. Ruppelt noted that there were concentra-
tions of cases in the Los Alamos-Albuquerque area, Oak Ridge,
White Sands, Strategic Air Command locations, ports, and
industrial sites.[226]

In other words, the new Project Grudge suggested that UFOs were sur-
veying our military and industrial sites, just the type of thing an adversary
might do if planning an attack. Nothing suspicious about that at all.

The sixth investigation was called Project Bear and ran from late 1951
to late 1954. From the results section:

The Project BEAR report was based on a statistical analysis of
UFO sightings and contained graphs showing their frequency
and distribution by time, date, location, shape, color, duration,
azimuth, and elevation. It concluded that all cases that had
enough data were resolved and easily explainable. The report
assessed that if more data were available on cases marked
unknown, most of these cases could be explained as well. It also
concluded that it was highly improbable that any of the cases
represented technology beyond their "present day scientific
knowledge."[227]

That's six reports all reaching the same conclusion: that there's probably nothing to the accounts.

But then came the UFO panic in Washington, DC, in the summer of
1952, which we covered in an earlier chapter. That came to the attention
of the CIA, which didn't seem to think the six reports that already had been
done were the definitive word on the subject.

That's how the seventh investigation came into being, the CIA Special
Study Group, which was established in 1952. (The report doesn't provide a
date of conclusion.) From the results section:

The Study Group assessed that 90 percent of the reports were
explainable and the other 10 percent amounted to "incredible"
claims but rejected the notion that they represented Soviet or
extraterrestrial technology. The group also studied Soviet press
and found no reports of UFOs, leading the group to assume

that the Soviets were deliberately suppressing such reports. The Study Group also believed that the Soviets could use reports of UFOs to create hysteria in the United States or overload the US early-warning system.

- In December of 1952, H. Marshall Chadwell, Assistant Director of OSI [Office of Special Investigations], briefed the Director of Central Intelligence (DCI), Walter Bedell Smith, on the subject of UFOs. Chadwell urged immediate action because he was convinced that "something was going on that must have immediate attention," and that "sightings of unexplained objects at great altitudes and traveling at high speeds in the vicinity of major US defense installations are of such nature that they are not attributable to natural phenomena or known types of aerial vehicles."[228]

While the AARO report of March 2024 has received strong criticism from the UFO community, in many instances it appears to us that the investigators were trying their best to prepare an accurate report, given the restraints under which they were operating. Namely, they could review historical documents already released and talk with individuals, both whistleblowers and alleged members of these Special Access Programs, but they lacked the ability to subpoena anyone or force the disclosure of previously classified documents.

With all of that being said, we consider it something of a victory that the report mentions the opinion of H. Marshall Chadwell, the assistant director of the Office of Special Investigations for the CIA, who was alarmed by the reports he read and did not dismiss the sightings described as natural phenomena.

This isn't to say he was correct. But, at a minimum, it means that a highly qualified individual working at the top reaches of the intelligence community believed this was a matter that we did not understand, and the case could not be closed.

Which in turn led Chadwell to sponsor the eighth UFO investigation, the Robertson Panel, headed by a leading scientist from the California Institute of Technology, H. P. Robertson. The panel utilized experts in "nuclear physics, high-energy physics, radar, electronics, and geophysics."[229] From the AARO report:

> The panel reviewed all USAF data and concluded that most reports had ordinary explanations. The panel unanimously concluded that there was no evidence of a direct threat to national security for UFOs or that they were of extraterrestrial origin.
>
> - The panel was, however, concerned with the outbreak of mass hysteria and how the Soviets could exploit it. They recommended the USG use various channels to debunk UFO reports and suggesting monitoring domestic UFO enthusiast organizations.[230]

A couple of items stand out from the AARO report as a whole. We're getting a consistent pattern of the government saying that UFOs do not pose a threat to national security, that most can be explained, but there's a growing concern with public reaction, a quiet effort by the government to monitor UFO groups, and a public effort by the government to debunk UFOs. One can't help but notice there's an inherent contradiction between saying that most reports have "ordinary explanations" and saying that there's an active program of "debunking" such reports.

This led to the ninth government investigation, called the Durant Report, which simply summarized the findings of the Robertson Panel, with the addition of material in a report drafted by CIA officer Frederick Durant and presented to the CIA's assistant director of the Office of Special Investigations.[231]

The tenth government investigation of UFOs was the largest and longest, Project Blue Book. It began in March of 1952 and ran until December of 1969. The investigation created a classification system for UFOs: "identified," "insufficient data," and "unidentified."[232] The report also adds great clarity as to those factors that might result in the false identification of a UFO:

- **Astronomical sightings:** These consisted of bright stars, planets, comets, fireballs, meteors, auroral streamers, and other celestial bodies. When observed through haze, light fog, moving clouds, or other obscurations or unusual conditions, the planets—including Venus, Jupiter and Mars—were often reported as UFOs.

- **Balloons:** These included weather balloons, radiosondes, and large research balloons with diameters up to 300 feet, which together accounted for several thousand cases. Balloons were released daily from military and civilian airports, weather stations, and research activities. Reflection of the sun on balloons at dawn and sunset sometimes produced strange effects which led to many UFO reports. Large balloons can move at speeds of over 100 miles per hour when in high-altitude wind streams.

- **Aircraft:** According to Project BLUE BOOK, various aircraft accounted for another major source of UFO reports; particularly during adverse weather conditions. The staff noted that when observed at high altitudes and at a distance, the reflection of the sun on aircrafts' surfaces can make them appear as "disc" or "rocket shaped." They also noted that vapor or condensation trails from jet aircraft will sometimes appear to glow fiery red or orange when reflecting sunlight.

- **Afterburners:** Bright afterburner flames from jet aircraft were often reported as UFOs since they could be seen from great distances when the aircraft was not visible.

- Other UFO researchers included **stellar mirages, satellites, missiles, reflections, searchlights, birds, kites, false radar indications, fireworks, flares,** and confirmed **hoaxes.**[233]

The overall impression from this information is that misidentification is much more likely when the object is seen from a distance. In the results section, the report says, "Of the 12,618 sightings in Project BLUE BOOK's holdings, 701 were categorized as unidentified and never solved."[234]

The eleventh governmental investigation, which began in 1964, was the CIA Evaluation of UFOs. It was prompted by a request from the director of the CIA, John McCone. The report says, "The CIA's scientific division officially acquired UFO-sighting case information from the director of the

National Investigations Committee on Aerial Phenomena (NICAP), a private organization founded in 1956."[235] This is interesting, in that the CIA was now working with a civilian UFO program by acquiring some of their records at the same time as the government was actively debunking UFO reports and groups.

The twelfth investigation, the O'Brien Committee, was established in 1964 and run by Brian O'Brien, a member of the USAF Scientific Advisory Board. It included Carl Sagan, a prominent astronomer from Cornell University.[236] In the 1970s, Sagan became perhaps the most recognizable face of science for his books and television appearances as the host of the show *Cosmos*.

> The committee's report stated that UFOs did not threaten U.S. national security and that it could find no case which represented technological or scientific advances outside of a terrestrial framework. The committee's primary recommendation was that this topic merited intensive academic research and that a top university should lead the study.[237]

The only difference here is the idea that the study should be led by a "top university."

The thirteenth investigation into UFOs, the Condon Report, was started in April of 1968 and run by Edward U. Condon, a physicist and former director of the National Bureau of Standards. It was funded by a $325,000 USAF grant to the University of Colorado. At least the Condon Report has an internal logic to it, although it's curious that nobody in the government seemed to pay much attention to it.

> The panel's report stated that: "Our general conclusion is that nothing has come from the study of UFOs in the past 21 years that has added to scientific knowledge. Careful consideration of the record as it is available to us leads us to conclude that further extensive study of UFOs probably cannot be justified in the expectation that science will be advanced thereby." The panel cautioned against support for scientific papers on this topic and recommended that teachers should not give credit to students for reading UFO literature and materials.[238]

Now we understand the genuine threat to America. It's not aliens or runaway government spending, but teachers giving credit to students when they read a UFO book. (Mazzola interviewed one of the surviving members of the Condon committee, Dr. Bob Wood, who claimed that when he earnestly tried to investigate the significant UFO cases, Edward Condon tried to get Wood removed from the committee. Mazzola's investigation of both the Condon Report and the Robertson panel have convinced him that these investigations were deliberately set up by the CIA to obfuscate the truth about UFOs.)[239]

It seems that people in government didn't appear to accept the Condon Report's recommendation that the subject shouldn't be studied, because the very next thing that happened was another investigation, the fourteenth, the National Academy of Sciences Assessment of the Condon Report, conducted in late 1968.

> After the Condon Report was criticized by some scientists—including Project BLUE BOOK's Dr. Hynek—a panel of the National Academy of Sciences (NAS) was tasked in late 1968 to examine the rigor, methodology, and conclusions of the Condon Report.[240]

While much of the country's attention in the late 1960s and early 1970s was taken up by controversy over the Vietnam War, civil rights, and the Watergate scandals, the UFO issue eventually popped up in the presidential race of 1976, as the Democratic nominee had seen a UFO and was interested in the alien phenomenon. That person, Jimmy Carter, went on to win the election, and soon after wanted to make inquiries into that phenomenon.

> Dr. Frank Press, Science Advisor to President Jimmy Carter, sent a letter to Dr. Robert Frosch, NASA Administrator, on July 21, 1977, suggesting that a panel be formed by NASA to see if there had been any new significant findings since the Condon Report.

> Five months later, NASA responded by stating that it was not warranted "to establish a research activity in this area or to any new significant findings on UFOs since the Condon Report."[241]

But the five months' lag time seemed to indicate a federal bureaucracy that didn't have much interest in what the citizens or their representatives wanted, which is perhaps why, during the ensuing Reagan and Bush presidencies, there were no official UFO investigations.

Instead, the American public had to wait for another Democratic president, Bill Clinton. Under his administration, there were four significant government attempts to obtain more information about UFOs, specifically revolving around the alleged Roswell crash of 1947, which occurred after the publication of several books, including that of Philip Corso, *The Day After Roswell*.

The fifteenth investigation into UFOs occurred between 1992 and 2001 and is titled "President Clinton and Chief of Staff [John] Podesta Inquire about Roswell[...]." The "investigation" seems to have been an informal series of requests. Per the report:

> According to press reports, President Clinton tasked former National Security Advisor Sandy Berger to determine if the USG held aliens or alien technology. President Clinton said, "As far as I know, an alien spacecraft did not crash in Roswell, New Mexico in 1947...If the USAF did recover alien bodies, they didn't tell me about it...and I want to know."[242]

When we read how government agencies treated both Carter and Clinton when they asked questions about UFOs, it seems to us that somebody other than the president is in charge of knowing the truth about these objects.

Perhaps we're reading between the lines, but Clinton appears to be saying that nobody definitively told him there were no alien bodies or technology in our possession; they simply refused to answer his inquiries.

The 1995 report from the sixteenth inquiry into UFOs is titled "The Roswell Report: Fact Versus Fiction in the New Mexico Desert." From the AARO report of 2024:

> The USAF conducted a systematic search of numerous archives and records centers in support of GAO's audit of Roswell. As part of this review, the USAF also interviewed numerous people who may have had knowledge of the events. Secretary of the Air Force Sheila E. Windall released them from any secu-

rity obligations that may have restricted the sharing of information.... The report stated that the USAF's research did not locate or develop any information that indicated the "Roswell Incident" was a UFO event, nor was there any "coverup" by the USG. Rather, the materials recovered near Roswell were consistent with a balloon of the type used in the then-classified Project Mogul.[243]

One might think of this report as yet another government denial, but the difference is that it adds a specific detail: Project Mogul, a top-secret weather balloon project of the United States government.

The seventeenth investigation by the government into UFOs was titled "The GAO Roswell Report":

The GAO's 1995 report on the results of its investigation found that the U.S. Army Air Force regulations required that air accident reports be maintained permanently. Four air accidents were reported by the Army Air Force in New Mexico during July 1947. All involved military aircraft and occurred after July 8, 1947—the date the RAAF public information office first reported the crash and recovery of a "flying disc" near Roswell. The military reported no air accidents in New Mexico that month. USAF officials reported to GAO that there was no requirement to prepare a report on the crash of a balloon in 1947.[244]

This is where things start to get confusing in this report. We'll concede that there can be some initial confusion trying to create a credible timeline of events from four decades prior, but this should have been a relatively simple matter.

First, we're given a hard date for the public announcement of a crashed "flying disc," and that date is July 8, 1947. We can safely assume the "crash" happened within a few days of that date, so the suggested date of July 3 or July 4 for it makes a good deal of sense, allowing a few days for learning, for investigating, and for recovering the debris. It must be admitted, though, that in those few days, nothing the Army Air Force personnel saw convinced them it was a weather balloon, because they initially reported it as a crashed flying disc.

Second, we're told that there were no "air accidents" reported around that time, although it's unclear whether a balloon mishap would have been reported as an air accident.

What is clear from the third Roswell report is the time frame in which the event happened.

This leads us to the eighteenth government investigation and its 1997 report, "The Roswell Report: Case Closed." This fourth report on Roswell seems to invalidate the previous reports' claims about the Roswell crash. From the 2024 AARO report:

- The USAF subsequently published a follow-on report in 1997, *The Roswell Report: Case Closed*, with additional materials and analysis which supported its conclusion that the debris recovered near Roswell was from the U.S. Army Air Force's balloon-borne program.

- The alleged "alien" bodies reported by some in the New Mexico desert were test dummies that were carried aloft by US Army Air Force high-altitude balloons for scientific research.

- Reports of military units that allegedly recovered a flying saucer and its "crew" were descriptions of Air Force personnel engaged in the dummy recovery operations. Claims of "alien bodies" at the Roswell Army Air Force (RAAF) hospital were most likely the result of the conflation of two separate incidents: a 1956 KC-97 aircraft accident in which 11 Air Force members lost their lives; and a 1959 manned balloon mishap in which two Air Force pilots were injured.[245]

The government sounds like a middle school student trying to explain why he didn't do his homework. First, it's a regular weather balloon, then it's a supersecret weather balloon, then people misidentify crash test dummies attached to balloons, then people are misremembering incidents that happened nine and twelve years later.

This led to the nineteenth and twentieth government investigations, the Advanced Aerospace Weapons Systems Application Program (AAWSAP) and the Advanced Aerospace Threat Identification Program (AATIP). From the background section on these programs:

> At the direction of Senate Majority Leader Harry Reid (D-NV), the Defense Appropriations Acts of Fiscal Years 2008 and 2010 appropriated $22 million for the DIA [Defense Intelligence Agency] to assess long-term and over-the-horizon foreign advanced aerospace threats to the United States. In coordination with the Under Secretary of Defense for Intelligence, DIA established AAWSAP in 2009, which was also known as AATIP.[246]

Whatever one's political affiliations might be, it must be agreed that the Senate majority leader is in a better position to know what's true or false than a typical member of the public, or even most other congressional members.

Add to that the fact that Senator Reid was from the state of Nevada, home to the mysterious Area 51, where there have been reports of captured or reverse engineered alien spacecraft, and one can only conclude that the senator had been privy to information and reports that left him deeply troubled.

To us it indicates a strong belief that what they were researching was worthwhile, and that the claims of previous decades, from weather balloons to misidentifications of the planet Venus, weren't holding up to scrutiny. In fact, it seemed as if the members of AAWSAP/AATIP were interested in moving the conversation in a completely different direction.

- On June 24, 2009, Senator Reid sent a letter to then Deputy Secretary of Defense William Lynn III requesting that AAWSAP/AATIP be made a DoD Special Access Program. Deputy Secretary Lynn declined to do so based on the recommendation of then-Under Secretary of Defense for Intelligence, James R. Clapper, Jr., that such a designation was not justified.

- Just prior to DoD's cancellation of the program, the private sector organization proposed as a new line of effort to host a series of "intellectual debates" at academic institutes to influence the public debate, which included hiring supportive reporters and celebrity moderators. The goal of this proposed public relations campaign was to assume that "E.T. visitations are true" and avoid the "morass" about discussing "evidence." A stated goal of the proposal was to increase public interest in government "disclosure" around the "E.T. topic" and explore the consequences of disclosure on the public.[247]

What does it signify when the leader of the legislative branch of our government has a significantly different view regarding UFOs than one of the highest officials in the intelligence branch, who does not want to comply with Reid's request?

One imagines that if there was nothing to the subject, the conversation between James Clapper and Senator Reid could be handled in about five minutes. That's what should have happened if there was trust between the highest levels of our legislative branch and the intelligence agencies. However, nothing even remotely close to that happened.

If anything, Senator Reid was suggesting he's just as suspicious as any UFO enthusiast.

Similarly, the "private sector organization" (billionaire Robert Bigelow's company Bigelow Aerospace) had become so convinced as to the reality of UFOs that it wanted to move beyond the question of whether these craft exist and urge the government to release the information it was currently holding, as well as wanted to hold discussions to influence the public about such disclosures. It seems as if those in the government who wanted to push the narrative that UFOs do not consist of anything mysterious in our world were losing control of the conversation.

The termination of the AAWSAP/AATIP programs led to the twenty-first government investigation of UFOs, the Unidentified Aerial Phenomena Task Force (UAPTF), which lasted from August of 2020 until November of 2021. From the background section of the report:

> Deputy Secretary of Defense David L. Norquist approved the establishment of the UAPTF in August of 2020 [during the height of the COVID-19 lockdowns]. Under the cognizance of the Office of the Undersecretary of Defense for Intelligence and Security (USD(I&S)), the Department of the Navy was asked to

lead the task force. It was established to improve understanding of, and gain insight into, the nature and origins of UAP. The task force's mission was to detect, analyze, and catalog UAP that could potentially pose a threat to U.S. national security.[248]

Seventy-three years and twenty federal government investigations after the 1947 Roswell crash, the US military was still trying to determine whether UFOs posed a threat to US national security. That's a long time and a lot of research for the world's most powerful nation to still be confused about weather balloons, high-flying birds, swamp gas, and mass hysteria.

In June of 2021, the UAPTF released a preliminary assessment of its findings:

The preliminary assessment concluded that: (1) the limited amount of high-quality reporting on UAP hampers the ability to draw firm conclusions about their nature or intent; (2) in a limited number of incidents, UAP reportedly appeared to exhibit unusual flight characteristics; although those observations could be the result of sensor errors, spoofing, or observer misperception and require additional rigorous analysis; (3) there are probably multiple types of UAP requiring different explanations based on the range of appearances and behaviors described in the available reporting; (4) UAP may pose airspace safety issues and a challenge U.S. national security; and (5) consistent consolidation of reports from across the USG, standardized reporting, increased collection and analysis, and a streamlined process for screening all such reports against a broad range of relevant government data will allow for a more sophisticated analysis of UAP.[249]

Does it seem as if the government is making any progress from its investigations of UFOs?

First, it says we don't have a lot of "high-quality reporting."

Second, in a limited number of cases the reporting does seem to be of high quality, but the authors believe this may be due to "sensor errors, spoofing, or observer misperception and require additional rigorous analysis." Does it seem unusual to anybody else that these "sensor errors" happened at exactly the same time as trained observers observed something unusual? It seems to me that whether one does or does not have "high-quality report-ing," the decision will *never* be made in favor of UFOs.

Third, assuming the data is correct, there are "probably multiple types of UAP," as they seem to look and behave differently.

Fourth, there might be "airspace safety issues" and "a challenge to US national security."

Fifth and finally, the government needs better reporting and analysis.

When the UAPTF program terminated in November of 2021, the twenty-second and twenty-third government investigation groups into UFOs were formed: the Airborne Object Identification and Management Synchronization Group (AOIMSG) and the Airborne Object Identification and Management Executive Management Committee (AOIMEXEC). They seem like parodies of useless government programs. The results of the seven-month investigation by these groups are described as follows:

> The organization helped initiate synchronization of efforts across the Department and the broader USG to detect, identify, and attribute objects of interest in "Special Use Airspace," as well as to assess and mitigate any associated threats of safety of flights and national security. AOIMSG had not achieved initial operating capacity before subsequent legislation in the FY2022 NDAA resulted in it being renamed AARO and given an expanded mission set.[250]

In other words, one group talked to a few groups in the government, which delivered nothing of value, and the second group never even got started until it was renamed.

The twenty-fourth government investigation of UFOs, the Unidentified Aerial Phenomena Independent Study Team (UAPIST), was in existence from June of 2022 until September of 2023. From the report:

> NASA released its report in September 2023. The report focused on discovering the best data streams available and dis-coverable to resolve UAP cases. It did not focus on whether or not UAP were of extraterrestrial origin. NASA also established a UAP Research Director position.[251]

If we're to believe this account, seventy-six years after the Roswell crash, the government finally has a system to determine how to look at

UFOs, although it's not a single step closer to figuring out what they are or how they are being piloted.

And finally, we have the twenty-fifth government investigation effort into UFOs, in the form of the All-domain Anomaly Resolution Office (AARO), established on July 15, 2022. The group reviewed the findings of its own initial report from January of 2023:

- The report stated that there were a total of 510 UAP reports as of August 30, 2022. This included the 144 UAP reports covered during the 17 years of reporting included in the ODNI's preliminary assessment, as well as 247 new reports and 119 reports that subsequently were discovered or reported.

- The report also stated that UAP events continue to occur in restricted or sensitive airspace, highlighting possible concerns for safety or adversary collection activity.

- The AARO Director reported to Congress that the majority of cases in AARO's holdings have ordinary explanations and that AARO has not seen any evidence that any of these cases represent extraterrestrial technology.

- *Of all the reports that AARO investigated and analyzed, none represent extraterrestrial or off-world technology. A small percentage of cases have potentially anomalous characteristics. AARO has kept Congress fully and currently informed of its findings. AARO's research continues on these cases.*[252]

When one considers the four main points of the initial AARO report, one can't help but conclude that we're still in the same position we've been in for decades.

Only two possible conclusions come to my mind: either this phenomenon is currently beyond our capability to understand, or the government has figured it out but doesn't want to share its answer with the public.

We found section five the most interesting, as it deals with approximately thirty whistleblowers who alleged a series of events that fell into a primary and secondary narrative. From the report:

The primary narrative alleges that *the USG and industry part-ners are in possession of and are testing off-world technology that has been concealed from congressional oversight and the world since approximately 1964, and possibly since 1947, if the Roswell events are included.* The narrative asserts that this UAP program possesses as many as 12 extraterrestrial spacecraft.

- An AARO interviewee claimed in a thirdhand account that an organization was in possession of 12 spacecraft recovered from different crashes prior to 1970. Some of the craft allegedly were "intact." The interviewee also stated that the CIA had a partnership with the company that ended in 1989 and wanted all material returned to the CIA. AARO discovered no empirical evidence supporting these claims.

- An interviewee claimed that an organization was in possession of off-world material in 2009 and 2010. A separate interviewee stated they participated in negotiations to return the material to the USG. The same interviewee stated that a former named senior CIA official quashed the proposal to remove the material from the corporation.

- A separate interviewee claimed that circa 1999, a former, senior U.S. military officer told the interviewee that he touched the surface of an extraterrestrial spacecraft. The interviewee stated that the senior officer gave a detailed description of a craft floating in a building. The officer told the interviewee that approximately 150 individuals were working on the program and that the program was kept "outside of government" so the technology could remain proprietary.

- Two interviewees said they participated in an alleged White House-tasked UAP study in Northern Virginia sometime between 2004 and 2007. The study evaluated the impacts to society should the United States, Russia, or China disclose they had evidence of extraterrestrial beings or craft. One interviewee assumed these governments possessed such evidence. The study was conducted by approximately 12 participants who evaluated 64 different aspects of society, such as religion and financial markets, which could be impacted by such a disclosure. The study lasted one day, and the interviewee was not aware of any final report or to whom any report may have been delivered.

- Another interviewee claimed that in the 1990s he overheard electronic communication of a conversation between two military bases where scientists claimed "aliens" were present during specialized materials testing. The interviewee also reported that on another occasion in the 1990s he observed an "unidentified flying object" at a U.S. military facility. The interviewee described the object as exhibiting a peculiar flight pattern.

- An interviewee who is a former U.S. service member said that in 2009, while participating in a humanitarian and security mission in a foreign country, he encountered "U.S. Special Forces" loading containers onto a large extraterrestrial spacecraft.

- A separate interviewee said that a family member was part of an effort to reverse engineer an object assumed to be off-world technology in the 1980s. The engineers failed to reverse engineer the object and it was sent to a different facility for further evaluation.

- An interviewee pointed out to AARO the existence of an alleged leaked Special National Intelligence Estimate from 1961 as proof of the existence of UAP crashes. AARO obtained a copy of the document through open-source research and evaluated its authenticity.

- Some interviewees and public accounts underpin this storyline by claiming through second and thirdhand accounts that some NDAs may have been used to protect a "reverse engineering program of off-world technology." These accounts describe the NDAs as including "punishment by death" provisions should the signatory disclose information about the program. Some interviewees claimed "verbal" and written NDAs were administered in several instances.[253]

The allegations appear to be specific, and yet there is considerable uncertainty regarding specific dates and the quality of the information provided. Mazzola notes that the account of the 2009 "service member" (later identified as Michael Herrera) reports US Special Forces loading supplies onto an "extraterrestrial craft." Mazzola questions whether this description is meant as a subtle way to discount the possibility of man-made reverse-engineered craft.

In determining whether an account is likely to be true, one looks for corroboration, and to us, one of the most interesting corroborations is the claim by two individuals that they were part of a "White House-tasked UAP study" sometime between 2004 and 2007, which looked at various aspects of an extraterrestrial disclosure campaign.

Probably the most outlandish claim is by the former US service member who "said that in 2009, while participating in a humanitarian and security mission in a foreign country, he encountered 'US Special Forces' loading containers onto a large extraterrestrial spacecraft." While the report does not name this individual, the service member had already gone public with his claims. An article published on the *Daily Mail* website on June 9, 2023, named him as Marine veteran Michael Herrera, and he detailed his claims to the newspaper:

> After 14 years of silence, Herrera was emboldened by new UFO whistleblower protections and in April testified under oath to the government's UFO investigation team, the All-Domain Anomaly Resolution Office (AARO), as well as a Senate committee.
>
> He provided his unblemished four-year service record, and texts about the incident with an alleged fellow witness—who refused to talk, saying it was 'not worth my life or jeopardizing my family.'
>
> Peripheral aspects of his account were verified by DailyMail.com using military sources. But Herrera, 33, does not have documentation or photos of the incident itself.[254]

This article indicates that Herrera is a credible witness—a service member with a clean record—and is willing to take an oath before Congress that he's telling the truth. In addition, he seems to have provided evidence that other members of his team were concerned for their safety if they were to speak out about the experience. This is the account from the article about their actions after a 7.6 magnitude earthquake hit Sumatra on September 30, 2009:

> Around October 8 he and five marines were dropped off at a clearing in the northeastern part of the city by a CH-53 chopper and hiked 900ft up a ridge to take their positions for the incoming supply drop, Herrera said.
>
> It was then he spotted a strange object in the jungle on the other side of the hill.
>
> 'I could see something moving and rotating. It was changing colors between a very light matte gray to a very dark matte black,' he said. 'It stuck out like a sore thumb.'
>
> Oddly, he said, they had not been given radios, so instead of calling it in they edge down the hill, in-formation, to investigate, while Herrera snapped photos and video on his Panasonic camera.
>
> 'The thing was massive, the size of a football field,' Herrera told DailyMail.com.[255]

We have the mental picture of a regular soldier on what should have been a routine deployment, although there's the question of why he and the other marines didn't get radios. Was somebody in a position of authority aware of what they might encounter, and had taken steps to curtail information gathering?

> Herrera claimed that when he and five comrades got within 150 feet of the craft, they were ambushed by eight men wearing all-black camouflage, bullet-proof vests, wielding M4 rifles with high-end night vision attachments given to elite US troops.
>
> 'Who the f*** are you guys? What are you doing here?' two of them yelled with American accents, he said.

'They said we weren't supposed to be there, and that they could kill us.'

As the men continued to threaten them, took the marines' weapon, dumped their ammunition and scanned their military IDs, Herrera said he saw others loading 'large weapon cases' and other containers from modified Ford F350 trucks on a platform beneath the craft.

'When the last two trucks finished unloading and drove off, the lower part of the platform rose off the ground to about 10 feet and the craft lowered to meet it, and it came together into one piece,' he said.[256]

American forces in Indonesia using alien craft to smuggle weapons? It sounds like the plot of a *Men in Black* movie. And yet, one has to ask, what would possess a person to make a claim to congressional investigators, understanding the claim would follow him for the rest of his life, unless he believed what he was saying was true?

Claiming to have seen a UFO invites ridicule, despite the fact that many famous people, including two United States presidents, have reported seeing them. The Marine's account of what happened in Indonesia continues:

'It rose off the ground and a little past the trees, then shot off to our left towards the ocean at around 4,000 mph.

'We can't believe this is f***ing happening. From a dead stop, it didn't make any sound like a sonic boom, it didn't disturb the trees like rotor wash would. We could see coconuts on the trees and none of them were disturbed.'

Herrera said the eight unmarked soldiers gave them back their unloaded guns and marched them back over the hill, 'still telling us how they could kill us.'

'Once we got over the hill they told us to get the f*** away from here and don't look back.'[257]

This is certainly the most dramatic of all the congressional UFO accounts, as it suggests an active program of reverse-engineering these craft, as well as the creation of military units dedicated to maintaining secrecy around it.

Let's return to the March 2024 AARO report and the secondary narrative that was developed through interviews with approximately thirty whistleblowers.

> The other narrative is that *a cluster of UAP sightings that occurred in close proximity to U.S. nuclear facilities have resulted in the malfunctioning and destruction of nuclear missiles and a test reentry vehicle.* AARO interviewed five former USAF members who served in and around a U.S. intercontinental ballistic missile (ICBM) silos, at Malstrom, Ellsworth, Vandenberg, and Minot USAF bases between 1966 and 1977. Some of these individuals claim UAP sightings near the silos, while others claim UAP disruption to ICBM operations. Specifically, they said the ICBM launch control facilities went offline or experienced total power failure. Additionally, one interviewee and a USAF videographer claimed to have observed and recorded a UAP destroying an ICBM loaded with a "dummy" warhead, mid-flight. AARO is researching U.S. and adversarial activity related to these events, including any U.S. programs that tested defensive ballistic missile capabilities.[258]

The report goes on to say that the investigators have not been able to resolve these allegations but hope to in a future volume.

Are you saying to yourself, "What the hell?" Five Air Force guys testified to the fact that UFOs were messing with our nuclear missiles and control facilities, and AARO basically said, "We'll have to do some further investigation?"

The report then went through a breakdown of many of the other claims, with titles such as "Former CIA Official Involvement in Movement of Alleged Material Recovered from a UAP Crash Denied on the Record," "The 1961 Special National Intelligence Estimate on 'UFOs' Assessed Not to be Not Authentic," and "Aliens Observing Material Test a Likely Misunderstanding of an Authentic, Non-UAP Program Activity."[259]

First of all, do we believe a CIA official when he denies having done something?

Second, the authors do not provide any information as to why they believe the 1961 Special National Intelligence Estimate to be inauthentic.

And as for the third headline, "Aliens Observing Material Test a Likely Misunderstanding of an Authentic, Non-UAP Program Activity":

> AARO determined this account most likely amounted to a misunderstanding. The conversation likely referenced a test and evaluation unit that had a nickname with "alien" connotations at the specific installation mentioned. The nature of the test described by the interviewee closely matched the description of a specific materials test conveyed to AARO investigators.[260]

When we read this explanation, we thought, "Maybe." If a number of claims are made, you have to expect that some of them are likely to be mistaken. But then our minds went to other questions, such as, "Why is there a test and evaluation unit that has a nickname with alien connotations? And what material were they testing?" We understand that some of the answers to these questions might be protected for national security reasons, but then that simply ensures that there will always be questions.

Here are some other examples of statements that seem to raise more questions than provide answers. For example, in regard to "The UAP with Peculiar Characteristics Refers to an Authentic, Non-UAP-Related SAP," referring to a whistleblower claim, there is this explanation:

> AARO was able to correlate this account with an authentic USG program because the interviewee was able to provide a relatively precise time and location of the sighting which they observed exhibiting strange characteristics. At the time the interviewee said he observed the event, DoD was conducting tests of a platform protected by a SAP. The seemingly strange characteristics reported by the interviewee match closely with the platform's characteristics, which was being tested at a military facility in the time frame the interviewee was there. This program is not related in any way to the exploitation of off-world technology.[261]

Color us suspicious, but when a government report admits that an individual saw a craft that exhibited "strange characteristics" that could easily

Former Navy Fighter Pilot, Ryan Graves, Retired Air Force Major, David Grusch and Retired Navy Commander, David Fravor testifying before the House Oversight and Accountability Subcommittee on Unexplained Aerial Phenomena (UAP) on July 26, 2023. (Getty Images)

Legendary constitutional law attorney, Daniel Sheehan, was deeply involved in the July 2023 hearing and has been investigating the UFO issue since the Carter administration. (Getty Images)

Above: The explosion of the first nuclear bomb at the Trinity Site in the early morning hours of July 16, 1945, which ushered in the atomic age, and perhaps the current interest in our species by "non-human" intelligences. (Getty Images)

Left: Inspection of the awesome power of the nuclear bomb by Dr. Robert Oppenheimer and General Leslie Groves. (Getty Images)

Author Paola Harris with Jose Padilla, who on August 16, 1945, was nine years old, and with his friend, seven-year-old Remigio Baca, were riding their horses when they discovered a downed alien craft. (With permission by Paola Harris)

Left: The drawing by pilot Kenneth Arnold, of what he claimed he saw on June 24, 1947, just prior to the alleged Roswell Crash of July 3-4, 1947. Arnold is credited with being responsible for the expression "flying saucers," but he claimed this referred to the way they seemed to "skip" through the air, rather than their shape. (Instock Images)

Below: Picture of a modem B-2 "stealth" bomber in flight. Many question whether this is an example of reverse-engineered technology. (Shutterstock Images)

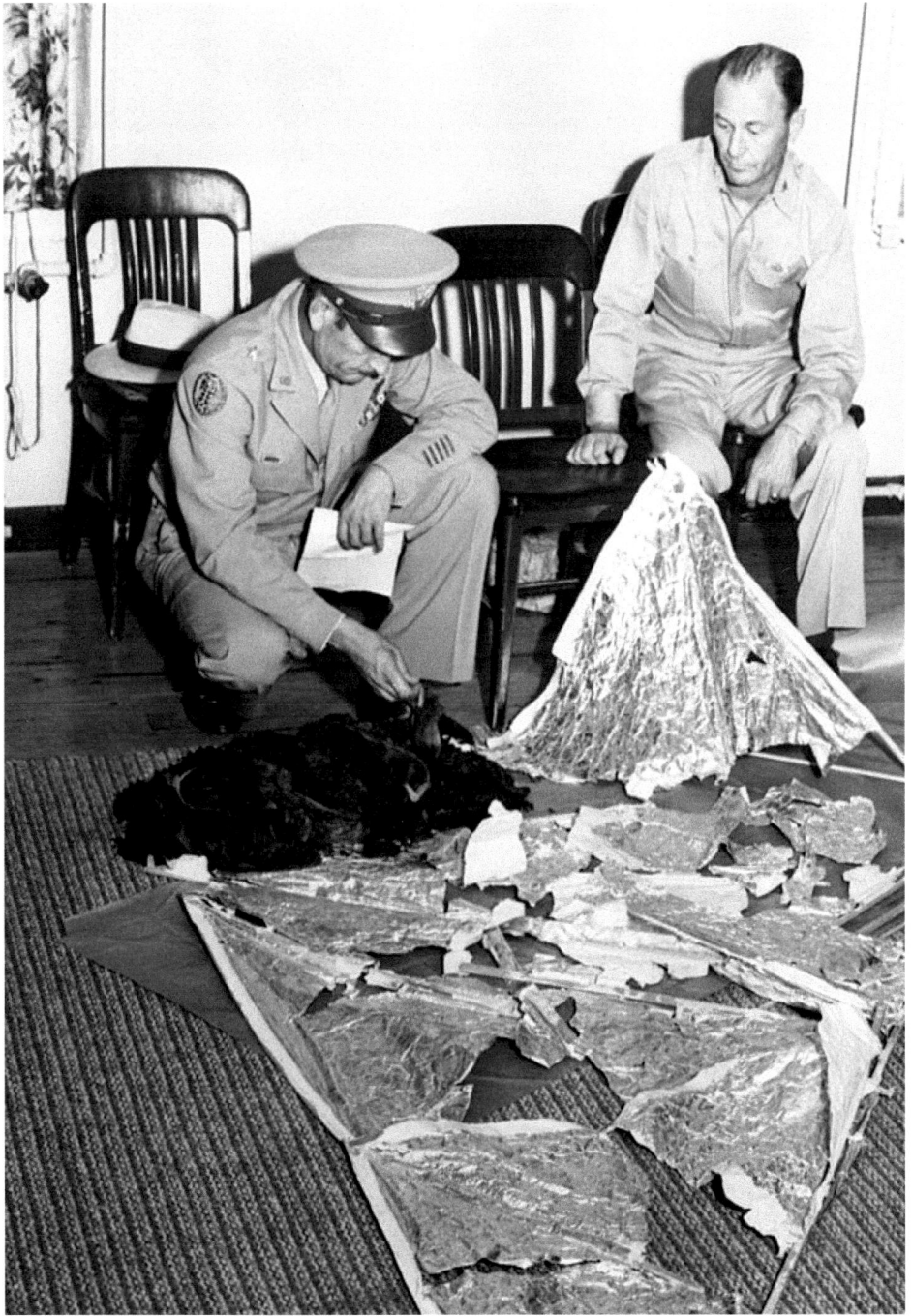

Brigadier General Roger Ramey and Colonel Thomas Dubose with what is claimed as wreckage of the Roswell craft (July 3-4, 1947), which the Army asserts is a downed weather balloon. Hi-tech scans of the message Ramey is holding seems to reveal the words "disk," "wreck," "aviators in the disk," and "sent out PR of weather balloons." (Getty Images)

Author Paola Harris with Colonel Philip Corso, who in his 1989 book *The Day After Roswell* described how in the early 1960s he participated in the study and reverse-engineering of the Roswell materials, as well as helping President Kennedy and Attorney-General Robert Kennedy plan the space program. (Used with permission of Paola Harris)

The scrambling of Air Force jets from Andrews Air Force Base on July 26, 1952, to pursue UFOs spotted over Washington D.C. (Mary Evans Picture Library/Pantheon/SuperStock)

UFO photographed near Holloman Air Force Base, New Mexico; never satisfactorily explained but a secret military device seems a possibility. (Mary Evans Picture Library/Pantheon/SuperStock)

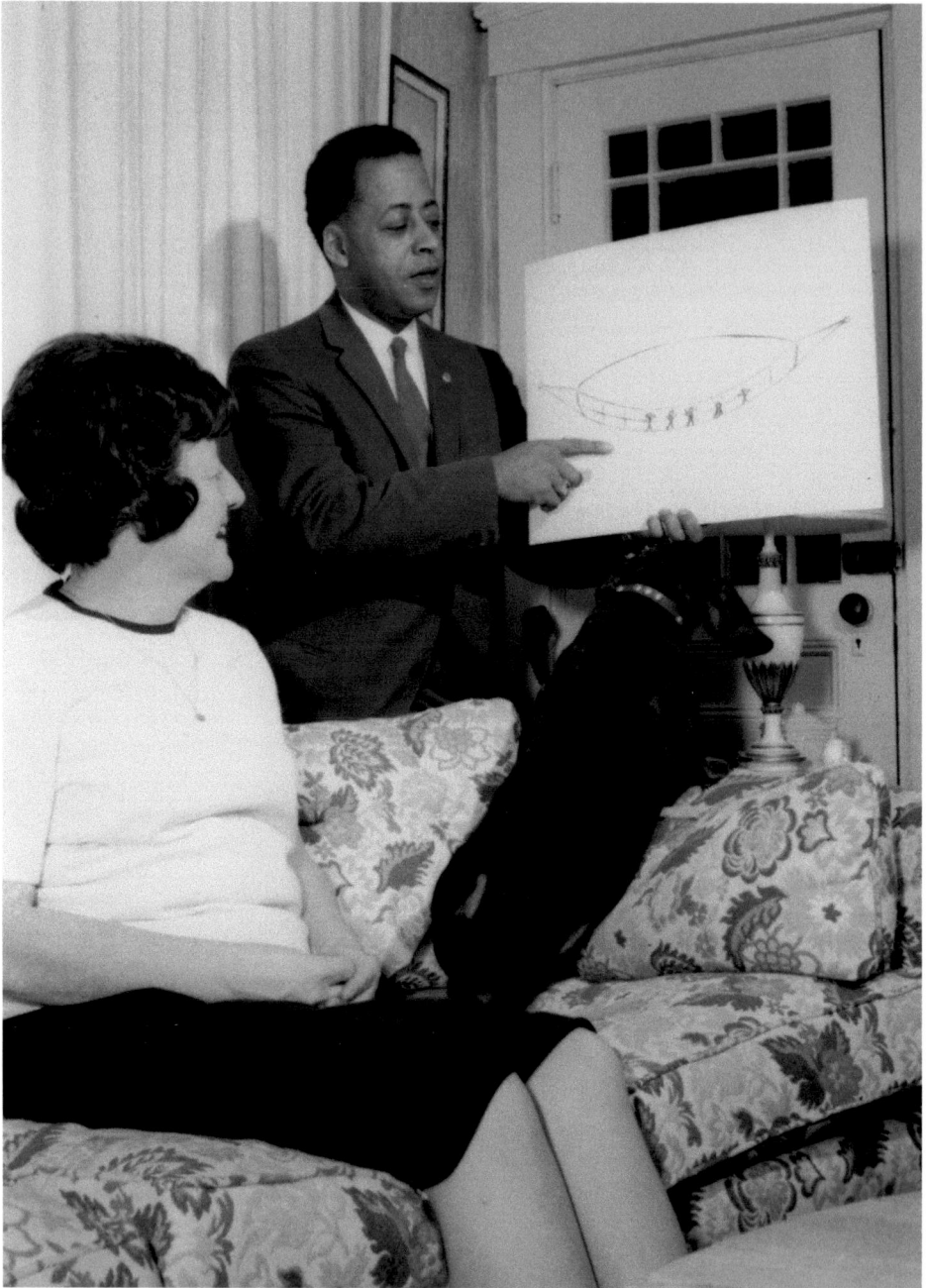

The first widely known UFO abduction case, that of Betty and Barney Hill, who claim that on the night of September 19–20, 1961, they were abducted while driving home. They wrote a bestselling book about the incident, which was published in 1966, and their story was turned into a television film by NBC in 1975, entitled *The UFO Incident*, starring James Earl Jones as Barney, and Estelle Parsons as his wife, Betty. (Getty Images)

7/21/86

Dear John:

Yes, I'm a believer in both categories. I feel everything is possible.

Many of our man-made UFOs are Un Funded Opportunities.

In both categories, there are a lot of kooks and charlatans — be cautious.

Best regards

Ben Rich

A 1986 letter from Ben Rich, head of Lockheed Martin, expressing his belief in both alien craft and man-made reproduction vehicles. (Courtesy of Michael Schratt)

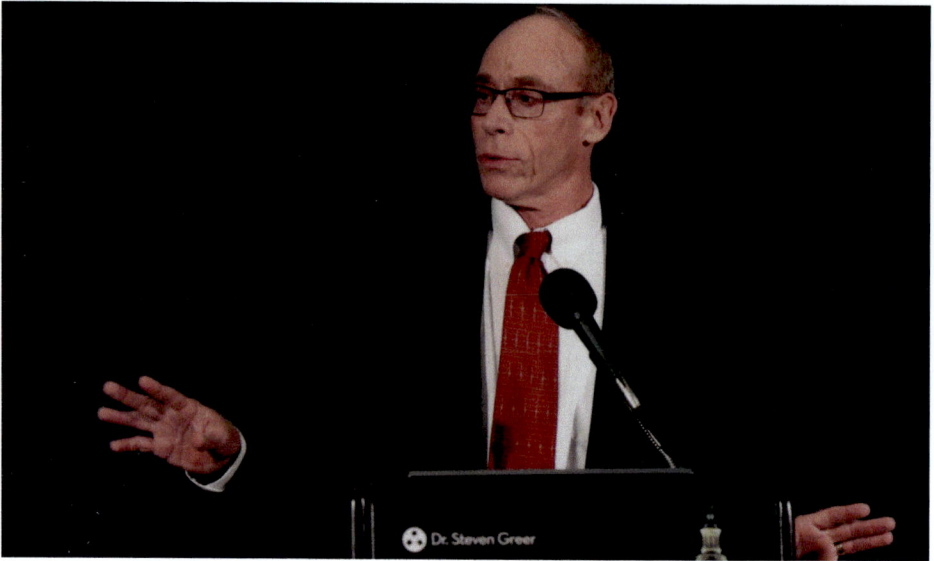

Dr. Steven Greer is arguably the leading figure in the UFO disclosure movement, working on the issue for more than thirty years, personally briefing President Bill Clinton's first CIA Director, Admiral James Woolsey, on a decades-long "Legacy" program of reverse engineering. (Getty Images)

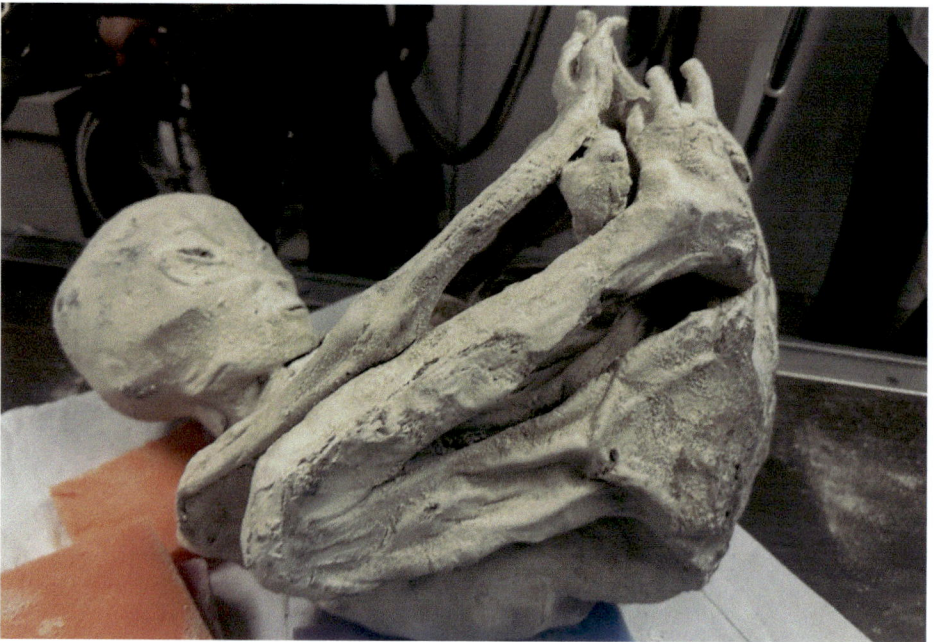

One of the largest controversies in modern Ufology is whether a series of desiccated mummies found near Nazca, Peru are evidence of an alien presence on our world. (Photo by Michael Mazzola)

Above: Dr. Jose Zalce examines an X-ray of a tridactyl skull. (Photo courtesy of Michael Mazzola)

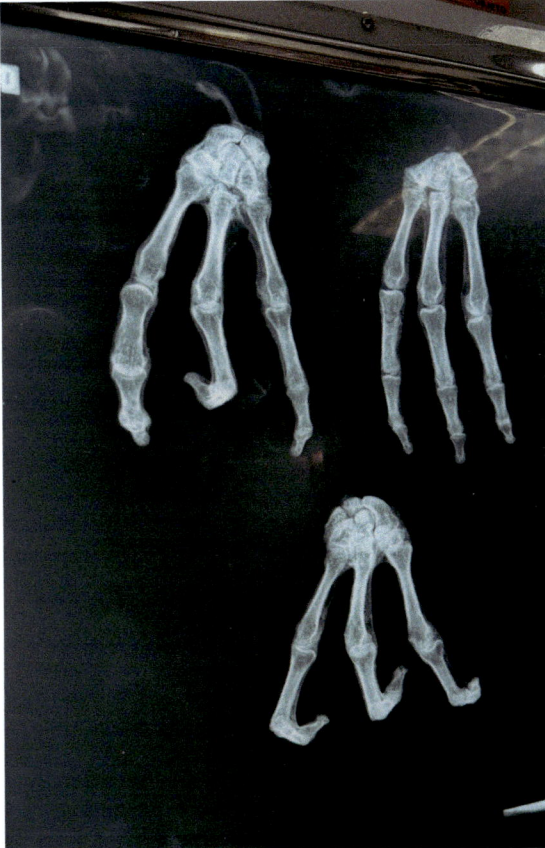

Left: X-ray of tridactyl hands excavated from a Peruvian grave site. (Photo courtesy of Michael Mazzola)

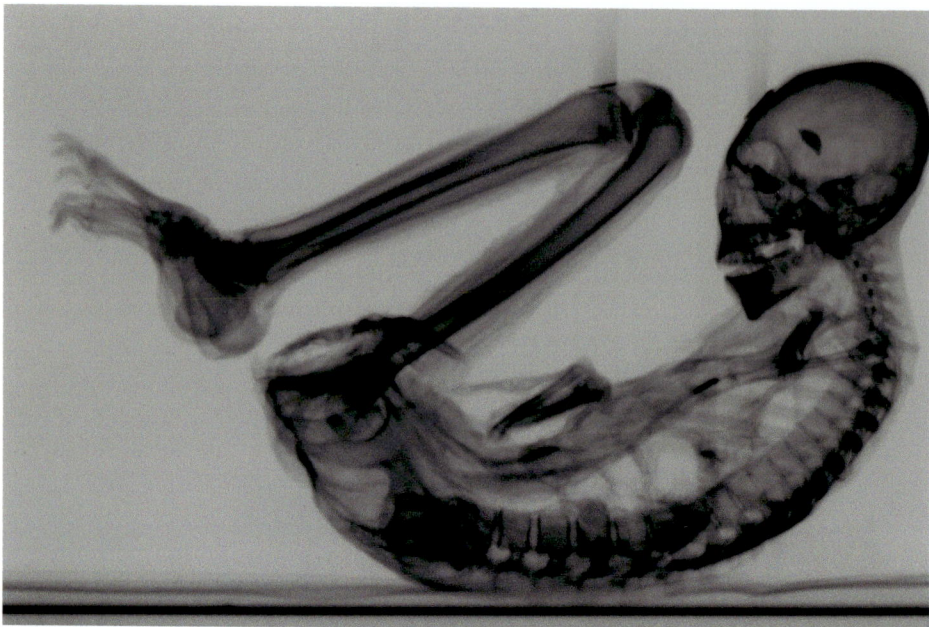

Above: X-ray of the Peruvian mummy known as "Montser-rat." (Photo courtesy of Michael Mazzola)

Left: Historic Peruvian tapestry depicting tridactyl humanoids. (Photo courtesy of Michael Mazzola)

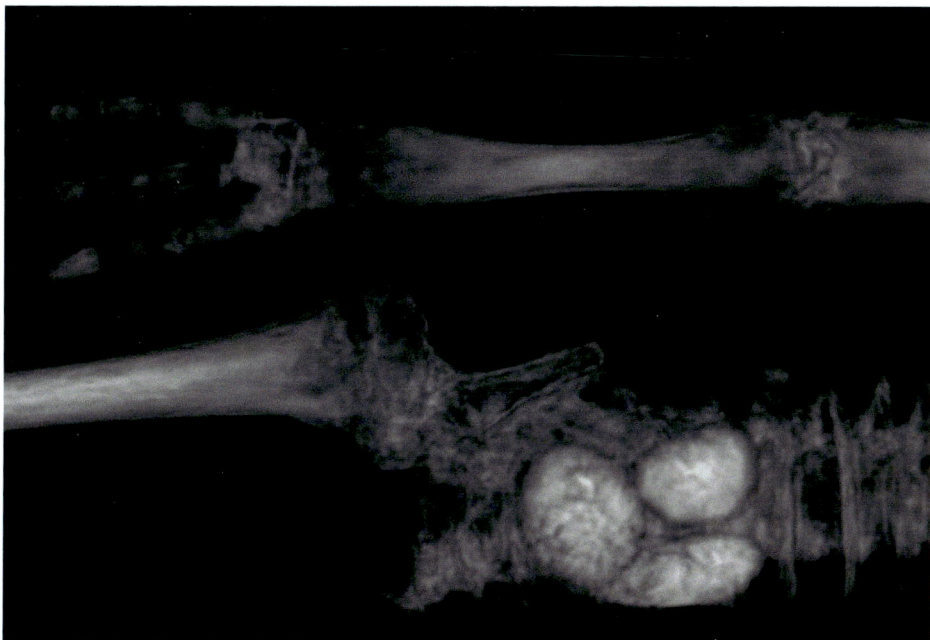

CT scan of a Peruvian mummy, showing three eggs and the three fingers of the hand. (Photo courtesy of Inkari Institute)

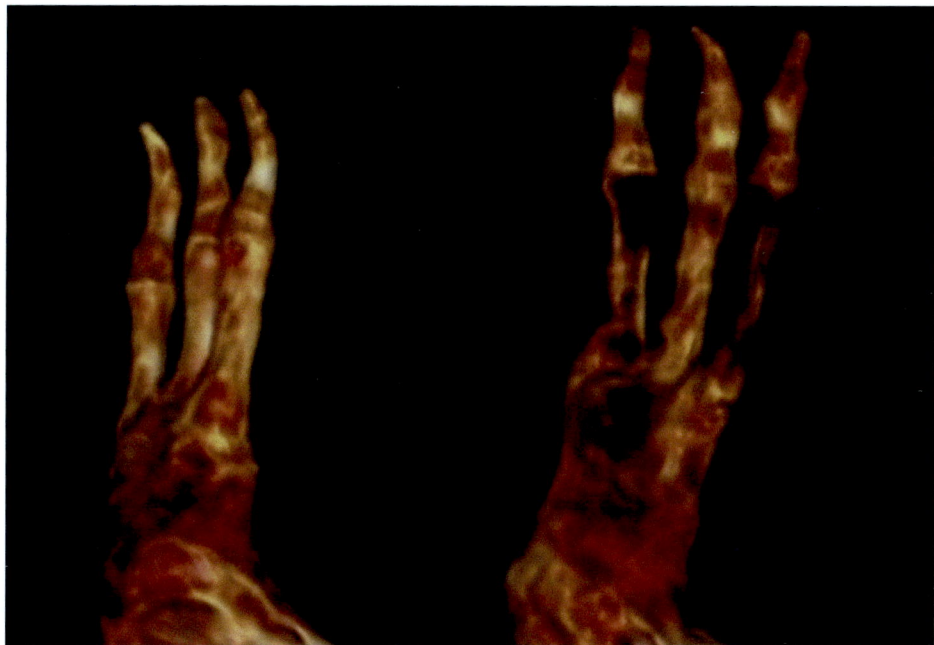

CT scan of the feet of a Peruvian mummy showing the three toes of the feet. (Photo courtesy of the Inkari Institute)

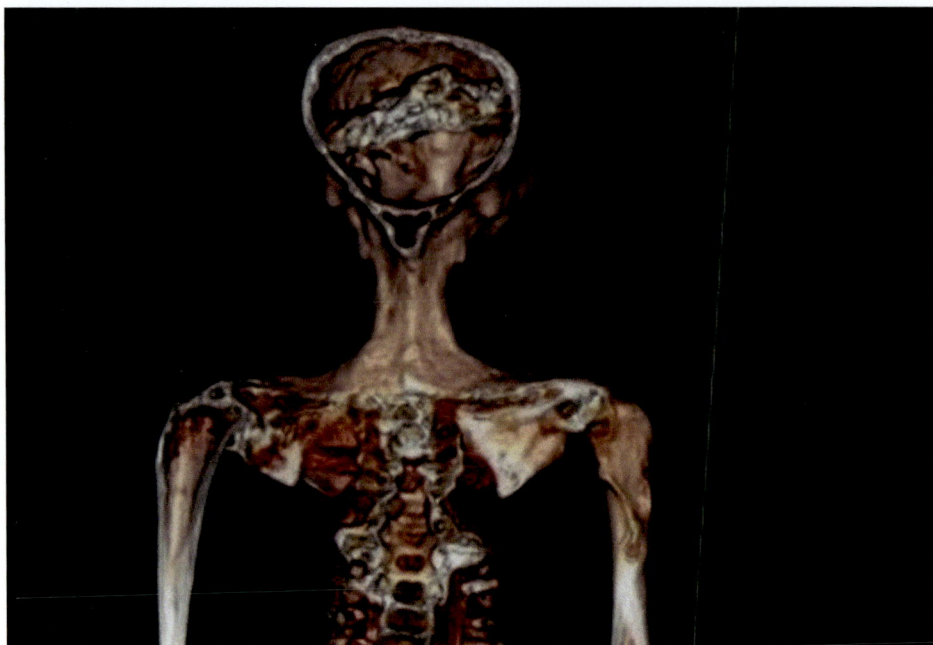

CT scan of a Peruvian mummy showing brain, spinal column, and joints of shoulders and arms. (Photo courtesy of the Inkari Institute)

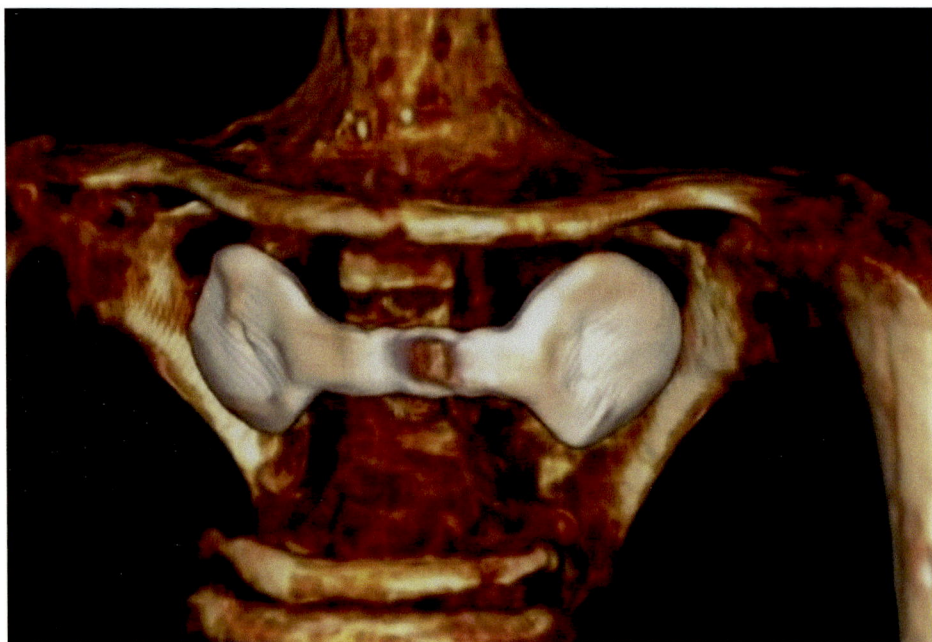

CT scan of a Peruvian mummy showing the mysterious metal implants on the creature. (Photo courtesy of the Inkari Institute)

Above: A picture of the mysterious "Buga" sphere. (Photo courtesy of Michael Mazzola)

Left: Dr. Steven Greer and Congressman Eric Burlison examine a large tridactyl hand in Mexico City on June 20, 2025. (Photo courtesy of Michael Mazzola)

Serena DC speaking at the tridactyl press conference at the Mondrian Hotel in West Hollywood in 2024. It has been suggested by scientists that the creatures be given the scientific name *homopan tridactylis* in recognition of their human, chimpanzee, and apparent alien DNA. Listening to the presentation are Michael Mazzola and Jaime Maussan. (Photo courtesy of Michael Mazzola)

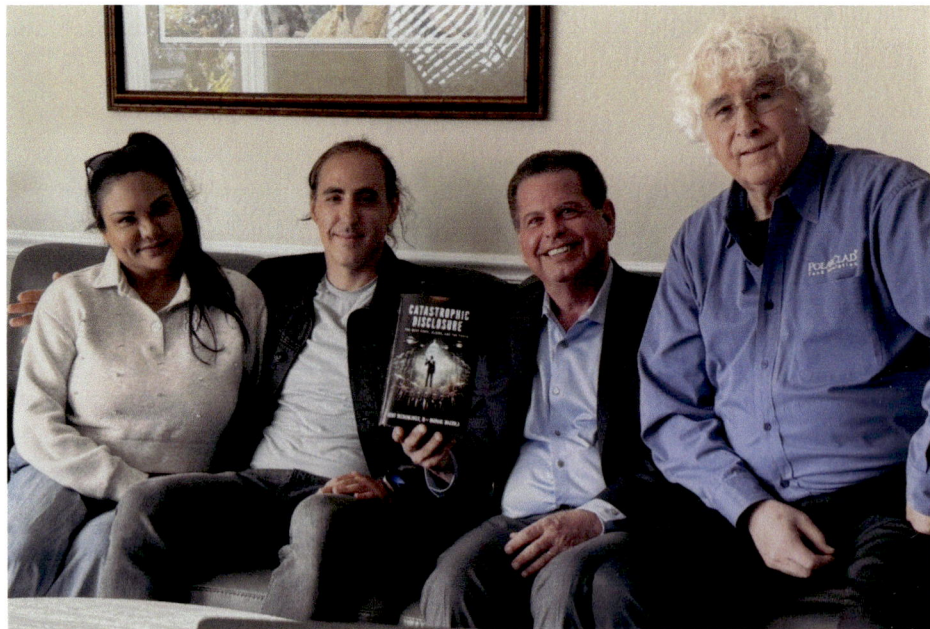

Serena DC, Michael Mazzola, Kent Heckenlively, and Daniel Sheehan at the filming of the documentary *This is Not a Hoax*, about the Nazca mummies. (Photo courtesy of Kent Heckenlively)

be mistaken for those of a UFO, but the craft didn't have anything to do with UFOs, we're going to need some more information before we believe it.

However, we're told the information is protected by a SAP (Special Access Program), which is how Dr. Steven Greer (as described in the set of interviews that began this book) has long claimed that this information is being hidden from the public. Point for Dr. Greer.

A similar problem plagued us when we read the heading for a different allegation, "Aerospace Companies Denied Involvement in Recovering Extraterrestrial Craft," and the following explanation:

> AARO met with high-ranking officials, including executives and chief technology officers of the named companies. All denied the existence of these programs and attested to the truthfulness of their statement on the record.[262]

We'd be interested to know how extensive the government's investigation of these companies was and whether they asked for supporting documentation.

It may sound like we're making light of AARO's work. But in an earlier section of the report, there is a discussion of nondisclosure agreements (NDAs), which many of the whistleblowers said they were required to sign. They also said that they were often told that if they were to reveal any of this information, they could be subject to the death penalty.

To our reading, the report doesn't specifically address the question of whether these whistleblowers were forced to sign such NDAs, or whether their claims of being threatened with death for revealing information are accurate. However, this is what the report says regarding the NDAs:

> In the conduct of this review, and to meet the direction of Section 1673 of the NDAA for FY 2023, AARO sent guidance and requests to DoD, IC [intelligence community] elements, DOE [Department of Energy], and DHS [Department of Homeland Security] to review and provide any NDAs pertaining to UAP (or its previous names). To date, AARO personnel have not discovered or been notified of any NDAs that contain information related to UAP. Also, apart from the standard NDA language contained in Title 18, Section 794 describing the death penalty or jail time for illegally disclosing informa-

tion related to the national defense, AARO has not discovered any NDAs containing threats to interviewees for disclosing UAP-specific information.

Historically, most if not all NDAs contained standard language stating that the death penalty can be applied for the crime of disclosing classified information. Title 18, Section 794, is references in typical NDAs in several places in relation to the transmission of classified information:

"Whoever, with intent or reason to believe that it is to be used to the injury of the United States or to the advantage of a foreign nation, communicates, delivers, or transmits...information relating to the national defense, shall be punished by death or by imprisonment for any term of years or for life..."[263]

We had to read that passage a few times to make sense of it, and you may need to do the same. But here's what we think they're saying.

AARO requested UAP-specific nondisclosure agreements from the Department of Defense, the intelligence community, the Department of Energy, and the Department of Homeland Security, and received nothing.

On the question of whether those insisting on NDAs can threaten death to those who reveal information, the answer is a resounding yes.

However, there seems to be an even greater problem. If the government is in possession of crashed UFOs and has reverse engineered them, that would be an enormous secret to conceal, and the government would deploy all necessary resources to protect that information.

The simplest way to achieve that objective would be to label alien craft or technology "experimental" under any NDA. If we were in the position of keeping this information secret, we'd never write in a nondisclosure agreement, "And whatever you do, don't talk about the alien vehicles, technologies, or bodies." We'd just say it was classified technology and leave it at that.

On the question of the claim by whistleblowers regarding these programs to reverse engineer alien technology, this is what the report says:

- *AARO concludes many of these programs represent authentic, current and former sensitive, national security programs, but none of these programs have been involved with capturing, recovering, or reverse engineering off-world technology or material.*[264]

This represents something of a profound shift regarding these witnesses. The report claims they saw something, but it wasn't alien technology. We can't help but wonder if this is maybe a partial disclosure that the government has something that could easily be mistaken for alien technology. In fact, this view is supported by a section of the report that details how AARO investigators obtained this information.

> AARO instituted a secure process for handling information to allow interviewees to come forward to provide their statements to AARO within secure facilities. AARO established a partnership with the Special Access Control offices for the DoD, IC, and DHS to review programs identified in interviews by name or description to determine if the programs correlated in time and location to historic SAP or Controlled Access Programs (CAP). This agreement details how interviewee claims concerning the names and descriptions of the alleged programs are handled, stored, and protected so that their veracity can be determined in a secure manner. A key part of this agreement is that AARO investigators have been granted full access to all pertinent sensitive USG programs.[265]

We have to give the AARO report credit for illuminating the architecture of information control regarding military secrets, even if it is still withholding information about UFOs and aliens. There are Special Access Programs and Controlled Access Programs, just as Greer has long asserted, and we must be careful of AARO's ultimate conclusion, because as the report states, investigators were granted full access only to "pertinent sensitive USG programs." The guardians of the secrets regarding alleged aliens could easily decide that a certain program is not pertinent to the AARO investigators, and the investigators would never know about it.

The AARO report also contains some information about a new program to study UFOs, KONA BLUE (the twenty-sixth governmental investigation), and shows that despite all the previous findings, some people working on this issue are not convinced by the government's pronouncements:

> KONA BLUE was brought to AARO's attention by interviewees who claimed that it was a sensitive DHS compartment to cover up the retrieval and exploitation of "non-human bio-

logics." KONA BLUE traces its origins to the DIA-managed AAWSAP/AATIP program, which was funded through a special appropriation and executed by its primary contractor, a private sector organization. DIA cancelled the program in 2012 due to lack of merit and the utility of the deliverables. As discussed in section IV of this report, while the official purpose of AAWSAP/AATIP was to conduct research into 12 areas of cutting edge science, the contractor team, and at least one supportive government program manager, also conducted UAP and paranormal research at a property owned by the private sector organization. [Author's note: the company was Bigelow Aerospace, and the location was Skinwalker Ranch in Utah.]

When DIA cancelled the program, its supporters proposed to DHS that they create and fund a new version of AAWSAP/AATIP under a SAP. This proposal, codenamed KONA BLUE, would restart UAP investigations, paranormal research (including alleged "human consciousness anomalies") and reverse-engineer any recovered off-world spacecraft that they hoped to acquire. This proposal gained some initial traction at DHS to the point where a Prospective Special Access Program (PSAP) was officially requested to stand up this program, but it was eventually rejected by DHS leadership for lacking merit. As demonstrated by the proposal package and by statements from the originator, Senators Lieberman and Reid, asked that the PSAP be established with the promise of additional funding....

It is critical to note than no extraterrestrial craft or bodies were ever collected—this material was only assumed to exist by KONA BLUE advocates and its anticipated contract performers. This was the same assumption made by the same individuals involved with the AAWSAP/ AATIP program.[266]

Here's how we interpret these claims from the AARO report.

When the two previous government investigations into UFOs (AAWSAP/ AATIP) ended, some of those participating in the effort believed the United States *did* have crashed alien vehicles, technology, and bodies, and they wanted to create a governmental structure to bring that information to the public.

Ponder that idea for a moment.

Those who'd been investigating UFOs at the request of the government didn't believe the government was telling the truth. And in addition, they got two powerful United States senators, Joe Lieberman, who in 2000 was the Democratic nominee for vice president, and Harry Reid, the majority leader of the Senate, to support their proposal.

And yet, somehow, the KONA BLUE proposal was not funded.

Or was it funded, as some claim, but under a cloak of secrecy?

⋏   ⋏   ⋏

The last part of the AARO report may be the most unconvincing, as it provides information about twenty-five formerly secret government programs that the authors claim may be at the heart of the misidentification of UFOs. I struggled with the question of how to present this information to you, and concluded that I should simply share directly from the report.

It is *your* document as a United States taxpayer, and I present the final six pages in their entirety. I have kept my comments sparse throughout those pages and saved my discussion for the end of the chapter. This is the government responding directly to the public, and I thought it best to give it to you in as unfiltered a version as possible.

After seventy-seven years of sightings, this is the government's official explanation of the UFO sightings:

MANHATTAN PROJECT (AUGUST 1942)

The U.S. effort to build an atomic bomb, the Manhattan Project, was named after the location of its initial offices in what became known as the Manhattan Engineer District at 270 Broadway, Manhattan, New York City. General Leslie R. Groves, head of the project, followed the custom of naming the U.S. Army Corps of Engineers' districts after the city in which they were located. The secrecy surrounding the Manhattan Project and the establishment of several other national laboratories, such as Los Alamos National Laboratories, Lawrence Livermore National Laboratory, Sandia

National Laboratories, Pacific Northwest National Laboratory, and Oak Ridge National Laboratory to support this effort probably contributed to the spike in reported UAP.

V-173/XF5U-1 "FLYING PANCAKE" (1942)

The V-173 aircraft flew for the first time on November 23, 1942. It was believed that maintaining a uniform airflow over the wingspan—or "pancake" fuselage—would allow the aircraft to take off and land at exceptionally low speeds without sacrificing high-speed performance qualities that would appeal to the USN for its fighter aircraft. The V-173 could take off vertically, had a circular wing 23.3 feet in diameter, and could almost hover. The XF5U-1's design was largely similar to the V-173. However, the USN cancelled the project in 1948 in favor of a switch to turbojet engines.

PROJECT MOGUL (1947–1949)

The U.S. Army Air Force Material Command operated Project Mogul between 1947 and 1949. The aim of this program was to secure intelligence on Soviet nuclear weapons testing and to provide an early warning mechanism for Soviet ballistic missiles. Specifically, Project Mogul scientists worked on developing high-altitude balloons that would carry sensors capable of detecting long-range sound waves from weapons tests or missiles traveling through the atmosphere. A crashed balloon associated with Project Mogul outside of Roswell, New Mexico, is assessed to be the source of early UFO claims. [Author's note: the first Soviet nuclear explosion took place on August 29, 1949, in Semipalatinsk, Kazakhstan, two years after the Roswell crash.]

PROJECT HIGH DIVE (1950S)

Project High Dive was a program that conducted tests on large balloons and used test dummies in its experimentation. The goal of this program was to research the effect on pilots when they ejected from aircraft, especially pilots' tolerance to deceleration from wind drag.

PROJECT AQUATONE/DRAGON LADY (1954)

President Eisenhower authorized Project Aquatone to develop the U-2 Dragon Lady, a high-altitude reconnaissance aircraft to collect intelligence on Soviet nuclear deployments. More than half of the UFO reports investigated in the 1950s and 1960s were assessed to be U.S. reconnaissance flights, according to a declassified CIA assessment on reconnaissance aircraft. The report noted that UFO reports would spike when the U-2 was in flight, especially from airline pilots to Air Traffic Control. At that time, commercial flights typically flew below 20,000 feet while the U-2 flew at 60,000 feet. The report noted that when commercial pilots were flying east to west, with the sun below the horizon, the sunlight would illuminate the U-2.

WS-117L/CORONA (LATE 1956)

In 1956, the USAF initiated the WS-117L satellite reconnaissance program equipped with a film-return vehicle. Following the launch of Sputnik, the Eisenhower Administration made this program a high priority. In February 1958, President Eisenhower decided the CIA would have the lead role in the program, called "CORONA," and that it would be jointly managed alongside the USAF. The CORONA program performed 140 launches between 1959 and 1972, with many returning film from space to the Earth for recovery.

VZ-9AV AVROCAR/PROJECT SILVER BUG (1958)

Canada initially led an effort to develop a supersonic, vertical takeoff and landing fighter-bomber in the early 1950s. A.V. Roe (Avro) Aircraft Limited (later Avro Canada) led the design for the concept, and this effort yielded the Avrocar, an aircraft with a circular shape that gave it a stereotypical "flying saucer" appearance. Canada pulled its support when the project became too expensive. The U.S. Army and the USAF took over the project in 1958 when Avro offered it to the USG, when it became known as "Project Silver Bug." Avro built two test vehicles that were designated as the VZ-9AV Avrocar, but the project was cancelled in December 1961 when the vehicle could not lift more than a few feet off the ground. Project Silver Bug was declassified in 1997. [Author's note: although the Avrocar did have a stereotypical flying saucer appearance, it couldn't fly more than a few feet off the ground, so I'm confused as to how this could account for UFO sightings.]

EXPLORER 1 (JANUARY 1958)

The United States launched its first satellite, Explorer 1, into space on January 31, 1958. Explorer 1 carried a cosmic ray detector and was designed, built, and operated by the NASA Jet Propulsion Laboratory.

OXCART/A-12/SR-71 (1958)

President Eisenhower approved the CIA-led program to develop a successor to the U-2 spy plane in 1958, which became fully operational in 1965. The U-2's successor, the A-12 OXCART sustained a speed of Mach 3.2 at 90,000-foot altitude. By the time the A-12 was deployed by the CIA in 1967, CORONA satellites were being used to collect imagery of denied areas with less provocation than aircraft overflights. In 1968, President Johnson ordered the retirement of the A-12 when it was replaced by the SR-71, which itself was a modified version of the A-12.

PROJECT MERCURY (1958-1963)

Project Mercury, America's first human space program made six flights. The objectives of the program were to orbit a manned spacecraft around Earth, investigate humans' ability to function in space, and recover astronauts and spacecraft safely.

TATTLETALE/GRAB (SEPTEMBER 1960)

The United States was the first nation to deliver a reconnaissance satellite to space. This electronic intelligence (ELINT) satellite was developed by the Naval Research Laboratory in early 1958 under the code name "TATTLETALE" with the mission of intercepting Soviet radar signals.

PROJECT GEMINI (1961-1966)

The Gemini program was a U.S. human spaceflight program that took place between the Mercury and Apollo programs. Similar to Project Mercury, Project Gemini spacecraft was launched using ballistic missiles that were designed to carry nuclear payloads. Project Gemini conducted 12 missions.

PROJECT APOLLO (1961-1972)

Project Apollo was a NASA human spaceflight program conducted after Project Mercury and Project Gemini. Project Apollo totaled 14 missions, 11 spaceflights, and 12 astronauts walking on the moon.

POPPY (1962-1977)

The successor to GRAB, Poppy was an ELINT satellite system developed by the National Research Laboratory that operated from 1962 to 1977 to collect Soviet radar emissions. A total of seven Poppy missions were launched between December 1962 and December 1971. The program was declassified in 2004.

GAMBIT (1963-1971)

The National Reconnaissance Office (NRO) launched its first high-resolution photo reconnaissance satellite system in 1963, which became known by its codename, Gambit. Two Gambit systems were developed: Gambit 1, initially launched in 1963, and Gambit 3, which was first launched in 1966. The Gambit 1 satellite's exposed film was returned to Earth in reentry vehicles, or "buckets," that separated from the satellite, fell through the atmosphere and descended by parachute until obtained by USAF aircraft at about 15,000 feet altitude. Gambit was declassified in 2011.

HEXAGON (1971-1986)

Similar to Gambit, Hexagon was an NRO photoreconnaissance satellite system. It was launched in 1971 to conduct wide-area searches of denied territory. From 1971-1986, 19 missions collected imagery over 877 million square miles of the Earth's surface. The Hexagon system was the last satellite employing film reentry vehicles. Hexagon was declassified in 2011.

SPACE TRANSPORTATION SYSTEM/SPACE SHUTTLE (1972–2011)

The Space Shuttle program was NASA's fourth human spaceflight program and was comprised of the first reusable spacecraft to carry humans into Earth's orbit. The space shuttle fleet—Columbia, Challenger, Discovery, Atlantis, and Endeavour—flew 135 missions, serviced the Hubble space telescope, and helped construct the International Space Station. The first shuttle launch, Columbia, was conducted on April 12, 1981.

HAVE BLUE/F117A NIGHTHAWK/TACIT BLUE (1975)

The Defense Advanced Research Project Agency (DARPA) oversaw the development of HAVE Blue in the mid-1970s, which was the first practical, combat-stealth aircraft. HAVE Blue completed its first test flight in 1977, and the success of this program led the USAF to later produce the F-117A Nighthawk, as well as the TACIT Blue aircraft. The HAVE Blue, F117A Nighthawk, and TACIT Blue programs laid the foundations for the later development of the B-2 stealth bomber.

ADVANCED TECHNOLOGY BOMBER/B-2 SPIRIT (1980)

The B-2 is a USAF low-observable stealth bomber capable of delivering conventional and nuclear payloads. It uses a combination of reduced infrared, acoustic, and electromagnetic signatures. It was first publicly displayed on November 22, 1988 in Palmdale, California and conducted its first flight on July 17, 1989. The first aircraft was delivered on December 17, 1993.

STRATEGIC DEFENSE INITIATIVE (MARCH 1983)

At the initiative of President Ronald Reagan, the Strategic Defense Initiative Organization was established in 1984 to explore a multi-layered strategic defense against ballistic missiles; this program involved research into space-based and ground-based systems including laser and interceptor missiles. This intensive research effort involved national laboratories and academia. Some of the technologies were determined to be years from development and funding was reduced. The program was ended in 1993 and was replaced by the Ballistic Missile Defense Organization.

ADVENT OF UNMANNED AERIAL VEHICLES 1980S-PRESENT

The research and development, flight testing, evaluation, deployment, and the operation of drones—Unmanned Aerial Systems (UAS), Unmanned Aerial Vehicles (UAV), Remotely Piloted Aircraft (RPA), and Remotely Piloted Aircraft Systems (RPAS)—almost certainly resulted in reported sightings of UAP. Some of these systems had a "saucer" or triangle-shaped appearance and were capable of loitering aloft.

The USG acquired and operated a number of systems for a range of missions including intelligence, surveillance, reconnaissance, and strike, among others. The below systems represent a sample of those that have been operationally deployed since the 1994 Bosnia conflict and subsequently employed in counterterrorism operations around the world. Since then, their form and use have spread to civil and commercial operations.

GNAT 750

The GNAT 750 was developed in the late 1980s by General Atomics Aeronautical Systems, Inc. The prototype served as the basis for a more advanced design under DARPA. It was first used in 1994 during the Bosnia conflict where satellites were not optimized to collect for extended times over such small areas and where the airspace was heavily defended by capable anti-aircraft systems.

PREDATOR

The Predator system, also built by General Atomics Aeronautical Systems, Inc., was based on the GNAT 750. It was initially a joint USN and U.S. Army project but transitioned to the USAF in 1996. It was known as the RQ-1. The system possessed synthetic aperture radar, electro-optical, and infrared sensors. It was used to support United Nations and North Atlantic

Treaty Organization efforts in Bosnia and was widely used in counter-terrorism operations. It became a platform with a wide array of technical capabilities that performed a variety of missions—such as close air support, combat search and rescue, precision strike, convoy/raid overwatch, target development and terminal air guidance. The USAF retired the fleet in 2018.

REAPER

General Atomics Aeronautical Systems, Inc., also built the MQ-9 Reaper—a newer, larger version of the MQ-1 Predator UAV. This platform is faster, equipped with more advanced sensors, can carry more munitions than the Predator, and can be easily tailored with a variety of mission-specific capabilities. The system requires a pilot to control the aircraft and an aircraft member to operate the sensors and weapons. It has an operational altitude of 50,000 feet. The aircraft is operated out of a variety of locations worldwide, including Creech Air Force Base (AFB) in Nevada.

THE DARK STAR

The RQ-3 Dark Star was a remote pilot-assisted stealth system intended to conduct reconnaissance missions in high-threat areas. Lockheed Martin, Boeing, and DARPA developed Dark Star in the mid-to-late 1990s. It never entered production, but the research conducted on Dark Star led to subsequent advances used on other platforms. Some observers asserted that Dark Star resembled a flying saucer with long narrow wings.

It was designed to be fully autonomous from its launch, mission engagement, and return. It used satellite links to transmit sensor data. The first prototype flew in 1996 but crashed a month later on its second flight. The system completed five test flights before DoD terminated the program in 1999 due to cost and instability problems. [Author's note: it's curious how there seems to be such a fascination with saucer-shaped aircraft, but the military just can't seem to get them to work. It makes you question why they keep trying.]

POLECAT

Lockheed Martin's Advanced Development Program organization, also known as "Skunk Works" developed the P-175 Polecat UAV to better understand the flight dynamics of tailless, bat-wing-shaped 90-foot wingspan high-altitude UAV, including the next generation of structural composite materials and configurations. The Polecat made its first flight in 2005, and Lockheed Martin disclosed its existence in the Farnborough Airshow in 2006.

SENTINEL

The RQ-170 Sentinel is a UAV developed by Lockheed Martin's Skunk Works for the USAF. It is a low observable platform with a variety of intelligence, surveillance, and reconnaissance payloads. The Sentinel is operated out of Creech AFB and the Tonopah Test Range in Nevada.

GLOBAL HAWK

Built by Northrop Grumman, the RQ-4 Global Hawk is the largest UAS in operation by the USAF. It can fly at 65,000 feet more than 34 hours and is also capable of loitering at 60,000 feet while monitoring almost 58,000 square miles. The aircraft is currently fielded in three different models.[267]

▲   ▲   ▲

There you have it, ladies and gentlemen: the twenty-five secret programs, finally revealed, which should prove to all beyond a shadow of doubt that there are no such things as UFOs.

"But wait," you might be saying to yourself, "how did the Manhattan Project generate UFO reports? They weren't doing anything in aviation. They were building a nuclear bomb." So, why is a government report, decades later, citing the Manhattan Project as the first likely cause of UFO reports?

It's really hard to confuse a mushroom cloud with a flying saucer.

Or how do flying cars which get no more than a few feet off the ground, or manned rocket launches that everybody knows about cause people to imagine they see discs in the sky? Or to convince military personnel that these objects are messing with our nuclear arsenals?

It's almost as if the government enjoys telling us the most outrageous possible lies.

The truth may well be out there, but the government has certainly gone out of its way to keep us from discovering it.

CHAPTER EIGHT

A CONSTITUTIONAL LAWYER TAKES ON UFO SECRECY

Daniel (Danny) Sheehan might be described as one of America's greatest civil rights attorneys, and as having had the most eclectic collection of legal cases of any lawyer in the country. One might be tempted to believe he is a character dreamed up by some eccentric Hollywood screenwriter with a passion for 1970s political dramas. But even after fifty years in the profession, he remains at the cutting edge of advocacy law, leading the charge for government disclosure of UFO information.

Perhaps Sheehan's path was laid down early, as evidenced by the fact that he graduated from Harvard Law School in 1970 and attended Harvard Divinity School. He completed the "Masters of Divinity" degree, and was preparing to enter the Ph.D. program, when he was recruited by the Jesuit National Headquarters as General Counsel to the Social Ministry Office. As a result, he never completed his planned doctorate. While in law school, Sheehan was one of the earliest members of the *Harvard Civil Rights-Civil Liberties Law Review*, and as an editor he developed the case that established the right of journalists to protect their confidential sources.

It was in his capacity as an editor there that he started collaborating with the law firm of Cahill, Gordon, Sonnett, Rheindle and Ohl, which at the time was the premier corporate litigation firm in America. At that time the Cahill firm represented NBC News. In 1970, as part of his work for the Cahill firm, Sheehan participated in the drafting of the Amicus Curae briefs for the "News Journalist Privilege" case before the U.S. Supreme Court, which was joined by ABC News, CBS News, *The New York Times*, and *The Washington Post*.[268] In 1971, the Cahill firm represented *The New York Times* in the Pentagon Papers case, which detailed U.S. military involvement in Vietnam from 1945 to 1968. The six to three decision by the court established the precedent that the government could not invoke prior restraint against the media relating to stories of past secret government operations for which no personnel were currently at risk.

Sheehan is also generally credited as the individual who uncovered the information that led to the revelation of the Iran-Contra scandal, which nearly brought down the Reagan administration in 1987.

However, Sheehan's greatest contribution to American democracy may be his current effort to lift the veil on government secrecy regarding UFOs.

▲ ▲ ▲

Sheehan's interest in the UFO issue dates back to the early days of the Carter administration in 1977.

In Sheehan's telling of Carter's interest in UFOs, on November 19, 1976, Carter, as the president-elect, sent word to then-CIA director George H. W. Bush to come down to Plains, Georgia, to personally brief him on the UFO issue and its potential relationship to any extraterrestrial civilization.

Bush reportedly told Carter that he had no reason to know such information. However, he offered Carter a deal: if Carter would keep Bush as CIA director, Bush might tell Carter what the CIA knew about the UFO issue.

Carter did not take that deal.

Instead, he asked the United States House Committee on Science, Space, and Technology to contact the Congressional Research Service to produce two major reports on the UFO issue, one on the possibilities of alien civilizations and one on whether such civilizations might be responsible for UFO sightings.

As a result, Sheehan was contacted by Marcia Smith, director of the Congressional Research Service's Science and Technology Division, to become special counsel for the investigation.

Sheehan was familiar with the UFO issue and had an attorney's sense of what information he most wanted to see; specifically, the several hundred unexplained cases of the long-running Project Blue Book investigation of UFOs.

Smith told Sheehan to report to the newly built Madison Building for the National Archives with two pieces of identification; he brought his driver's license and passport. In Sheehan's telling, the building was not even fully occupied at the time. Two men in suits were waiting for him at a door and asked for his identification, which he showed them. He was then told, "Go down the hallway to the right, and you'll see the elevator on the left, then go down to the basement. You'll see where to go."

Sheehan did, carrying his briefcase, and out of instinct opened it and took out a yellow legal pad, which he placed under his arm. When the doors opened, Sheehan saw a light down the hallway and started walking toward it. He eventually came to a door guarded by two other husky guys in suits, who asked to see his identification.

The guys checked the IDs, then told Sheehan he couldn't bring his briefcase into the room. Sheehan put the briefcase down but kept the legal pad under his arm. Amazingly, the two guys let him into the room with his legal pad and a pen in his pocket.

Sheehan found himself in a relatively small room, maybe twenty feet wide, and ten feet long, with three large foldout tables. On the tables there were light green government shoeboxes with little ties on them, as well as a microfiche machine. After checking over his shoulder, Sheehan opened one of the boxes and began looking through the canisters of microfiche film.

Most of the items he saw in the first reel of microfiche film were documents, which he quickly scanned. Not knowing how long he was going to be allowed in the room, Sheehan looked for photographs. There were no pictures in the first roll or the second, but in the third he struck gold: a series of black and white photographs of a crashed flying saucer.

In the photos, there was snow on the ground, and it appeared to be winter. The saucer was disc-shaped, about forty to fifty feet across, with a large dome on the top. As best as Sheehan could determine, the craft appeared to have hit a snow-covered field, plowing up a large ditch in its wake and finally coming to rest at a forty-five-degree angle against an embankment on a hill. The craft was surrounded by what appeared to be Air Force personnel in winter jackets with fur around the hood, and from the clothing and quality of the pictures, Sheehan estimated that the photos had been taken in the late 1940s or early 1950s.

In one of the pictures, Sheehan saw an individual with an old-time movie camera, the type with two large canisters on top of it. As Sheehan looked closer at one of the pictures, he saw there were symbols along the lower edge of the dome atop the saucer. He'd studied various languages and noted that they didn't appear to be Russian or Chinese writing or hiero-glyphics. He took out his yellow pad and sketched the symbols on the card-board backing of the yellow pad, hoping they wouldn't check all the way to the back. He recalls that he sketched about eight to ten of the symbols, then started to get nervous about what he'd done.

Sheehan quickly left the room, surprising the two guards posted out-side the door. This time, they noted his legal pad and asked Sheehan to hand it over to them. However, they simply gave flipped through the pages, not checking the cardboard backing, then handed it back to him. Sheehan picked up his briefcase and left to return to Jesuit headquarters.

At Jesuit headquarters, he showed the symbols to Father William Davis, his superior, as Sheehan was actually a candidate for the priesthood at the time. Davis took a look at the sketches, then reached down into a desk drawer and drew out an eight-and-a-half-inches-by-eleven-inches manilla clasped envelope, which he opened.

Davis showed Sheehan a clear picture of a UFO in flight.

"Where did you get this?" Sheehan asked.

"From my sister, Dody," said the priest. "Her husband, Mike, is the Chief Air Traffic Controller at the Seattle airport, and it was given to him by his best friend, [Sheehan couldn't remember the name], who was a pilot. He didn't want to get in trouble, so Dody said to give it to me because I was a Catholic priest."

Although Sheehan initially had been given access to secret files, his fur-ther UFO investigation efforts were fruitless.

In Sheehan's estimation, the lies told about UFOs should be considered in the same category as the lies about Vietnam, the lies about Watergate, and the lies about the Iran-Contra scandal.[269]

Whenever the government engages in sketchy behavior, its agents will lie about it.

⋏　⋏　⋏

On February 11, 2024, Sheehan gave a long address on the UFO issue to celebrate his newly founded New Paradigm Institute, dedicated to over-coming the government's concealment of UFO data.

He began by noting whistleblowers Luis Elizondo and David Grusch, as well as the efforts of Christoper Mellon, who was deputy assistant secretary for intelligence during the Clinton administration and the George W. Bush administration.

Sheehan considered Grusch's testimony to be extremely important, because he had gone on the record before Congress and made four extraordinary claims, which the intelligence agencies and congressional members would have to contend with:

First, the United States government was in possession of crashed extraterrestrial spaceships.

Second, the United States was in possession of bodies recovered from these spaceships, and DNA tests had shown they were nonhuman.

Third, the United States had been engaged in a decades-long UFO-retrieval program and reverse-engineering program for the purpose of delivering nuclear warheads within minutes into Russia and China.

Fourth, the Defense Department had allowed Grusch to testify before Congress as an official member of the Unidentified Aerial Phenomena Task Force.

The congressional response was remarkably bipartisan for a divided Washington.

All seventeen members of the United States Senate Intelligence Committee (eight Republicans and nine Democrats) unanimously agreed to a sixty-four-page draft bill that directed the military and intelligence services, as well as civilian defense contractors, to turn over all UFO materials to Congress. As Sheehan said in his speech:

> That means in plain English, give us the saucers back. You've got to turn over the saucers that you've recovered and the bodies that you've recovered. You have to turn them over and allow the National Archives to have these things. That means that the members of the Senate Intelligence Committee and the House Intelligence Committee who have the proper clearances can see these even though they remain classified. So, that is taking the information out of the exclusive hands of these agencies that have held it for so long and turning it over, in effect, to Congress. This is extraordinarily important from a Constitutional perspective, because it's the Congress of the

United States that are the elected representatives of all of our people. The people of the military and the intelligence agencies are not. Those people are not elected. They're hired people who work for the government.[270]

However, that effort was derailed by a few congressional members.

Whatever one makes of Sheehan's claims, there's no doubt that he's currently at the center of UFO disclosure efforts. In addition, for more than twenty years, Sheehan has been legal counsel to Dr. Steven Greer's Disclosure Project.

Sheehan then got to the heart of the matter: why the military and intelligence complexes don't want to reveal what they know about UFOs. He said it has to do with the development of a weapon called Prompt Global Strike—a missile that uses alien technology and can strike within the heart of China or Russia in two minutes.

"It's a complete first-strike weapon," Sheehan said. "It completely disassembles the entire cartography of the kind of geopolitical agreements that currently keep the world relatively safe. That's what they're doing. They don't want anybody to know we're doing that. What they're saying is, 'Well, we can't let anybody know we're doing that.' I say, 'If I know it, and the Russians and the Chinese know it, you're not hiding it from anybody but the American people, who aren't going to want you to do that.' What we need is a treaty with Russia and China and everybody else that nobody is allowed to use any of the technology from UFOs to make any kind of a weapons system. Period."[271]

Although Sheehan is doubtful about the chances of such an accord, he is amazingly optimistic about what lies ahead for the UFO disclosure effort.

⋏　⋏　⋏

The New Paradigm Institute, founded by Danny Sheehan, has an office right across from the offices of the House Intelligence Committee in the US Capitol, and those involved regularly brief lawmakers and government officials. As Sheehan explained in a public speech:

I'm sitting in closed-door meetings with the inspector general of the United States Defense Department and military officers in full uniforms, with Luis Elizondo, briefing them on the fact that there are extraterrestrial beings here.

They're going, "What? Right in my own territory and I don't know about it?"

"Yes, I'm afraid so," I tell them.

We have to understand what the repercussions of this really are. Virtually every one of our operational worldviews is based on the belief our species is the most intelligent, highly evolved, sentient beings in the whole universe. The only one.

The bottom line is that this is going to be an important reality, because people don't understand the underlying predicates for their worldviews. You have people saying, "Oh, that person's a liberal." or, "That person's a conservative." or maybe, "This other person is a reactionary." They throw all these terms around, but people don't understand it's all based on the same underlying worldview.... We're going to have to come to grips with the shortcomings of that worldview.[272]

On May 23 and May 24, 2024, coauthor Kent Heckenlively interviewed Danny Sheehan to more fully understand his efforts. He began by telling him the title of this book, *Catastrophic Disclosure*, and that we were focusing on the best information we could find from the past eight decades, since the alleged 1945 Trinity crash, paying special attention to official government reports. Heckenlively told Sheehan he was especially confused as to why the government has had at least twenty-five official investigations of UFOs if there's nothing to the phenomenon. His response:

The degree of obfuscation and the duplicity that's involved is a flag, in and of itself, of how seriously they want to keep this thing concealed and not talk about it. And they go from UAP, which they start out by referring to as "unidentified aerial phenomenon." Then they change it to "unidentified anomalous phenomena," and it starts to include ghosts and apparitions and Bigfoot and alleged portals—all this other kind of stuff, because they don't want to stay focused on UFOs, which is exactly what we're talking about.[273]

They talked for a few minutes about the March 2024 AARO report, and Heckenlively mentioned several of the shortcomings presented in the previous chapter. Sheehan agreed, and the discussion moved on to the whistleblowers and the information they'd been allowed to share.

> The recovery of bodies which are of nonhuman origin—we've got those pieces of information in front of us right now. Now they've steadfastly refused to provide any information whatsoever from those witnesses with face-to-face direct encounters with extraterrestrial beings.

> And they refuse to talk about any particular direct lines of communication that have been established between whoever these authorities are inside the Defense Department and the intelligence community with any of the one or more extraterrestrial species. That's like the third rail. You don't go near that.[274]

Heckenlively was genuinely enjoying this discussion, as it matched much of what he'd concluded. As much as there were several secret things coming to light from government reports, it didn't seem the entire truth was not being told. Sheehan began to drill down into how the intelligence agencies were still trying to lie to the American public.

> They're still trying to maintain the confusion of whether or not this is an extraplanetary or extraterrestrial vehicle, as distinct from an extradimensional vehicle. But it's because of the way in which they travel, whatever their means of propulsion is, that they mask and come out. They seem to suddenly appear, or materialize in front of people, and they can disappear. [Author's note: I took this to mean that Sheehan was saying the UFOs had some sort of cloaking device, which would be consistent with other reports that said some UFOs were visible on sensors but not to the naked eye.]

> And so, what they're doing is trying to obfuscate this further by saying, "Oh, this must be extradimensional," as distinct from just extraterrestrial. The extraterrestrial vehicles with organic beings are coming from another star system who have clearly

mastered superluminal [faster-than-light-speed] technology, and that's a bedrock issue, period[…].

I've talked to two people, face-to-face, who stood right there and watched as one of these UFO beings was being interviewed.[275]

What Sheehan was saying was mind-blowing but also along the rational lines of a legal investigation, following a logical chain of cause and effect.

If it becomes clear that the answers being given are obvious lies, one is free to speculate on what the truth may be. As a lawyer himself, Heckenlively understood an attorney begins an investigation in a spirit of skepticism. If the answers given don't make sense, that skepticism will deepen. Heckenlively mentioned to Sheehan that even when he gave a great deal of latitude to the government's explanations, much of it did not pass the smell test.

Heckenlively then asked for Sheehan's opinion of what's really going on behind the scenes. This was part of his response:

The most important thing that I have reason to believe is happening is that the so-called crash retrieval program may well be more than a passive retrieval program. There may be what they call a kinetic aspect to the program—that they're actively engaged in seeking out UFO vehicles and attempting to interfere with their navigation capacities to try to bring them down and recover them. And that's a very, very closely guarded secret.

But it's interesting that they have deigned to talk publicly about the crash retrieval program, which again is a kind of modified hangout position. That, oh, there's some people talking about the fact we have this kind of benign program, almost like a street sweeper operation going on. You can't just have wreckage lying around. You've got to go out and clean it up. Because if that's all they're doing, there are significant indications that there's a more kinetic aspect to the program recovering the craft.

The second thing is that there's a technology they've developed that is capable of finding and monitoring UFOs even when they're masked, when they haven't manifested in a material form that can be seen or picked up on regular radar or monitored in other ways.

They've got a program called Golden Dome that allows them to find, locate, and track UFOs in their still-cloaked manifestation, because they're apparently functioning in some vibrational frequency that is beyond the scope of normal radar and things. That's quite important.[276]

Heckenlively understood that this was an important point and wanted to make sure he fully understood it. In the study of UFOs, there are some who claim the craft originate from different dimensions. Sheehan did not accept this explanation, believing instead they were only concealing themselves from our perception, but still remaining in this dimension.

For clarification, Heckenlively asked Sheehan whether it was more appropriate to say the UFOs were phasing in and out of our dimension, or only in and out of our ability to detect them?

It's just the latter. There's light that's vibrating at a frequency above that which we can visually see with normal vision. And there's infrared light at a lower vibrational frequency; there's ultraviolet. And so, we have a fairly narrow band of the electromagnetic frequency that are light vibrations that we can see. And we've developed some various forms of technology that can probe into the higher-frequency vibrations and lower vibrations.

But there is another, much more subtle vibrational dimension that is phenomenological, and actually exists. So, it's not that they're coming in and out of existence. And it's not, I think, accurate to refer to them as coming from another dimension.[277]

Heckenlively then pivoted to what has frustrated him most in writing this book: the lack of a narrative in which he could be confident. It seemed to him that the entire field of UFO research was plagued by a puzzling set of facts, upon which people have made assumptions that might be wildly incorrect. He mentioned the August 1945 Trinity crash, the Maury Island incident, Kenneth Arnold's sighting in June of 1947, and the July 1947 Roswell crash. In three of these first four significant incidents of modern

UFO sightings, the craft seemed to be dealing with some kind of mechanical malfunction, suggesting to me that the effort to get to our world involves a substantial degree of risk.

Heckenlively told Sheehan he was reminded of the early European contact with the Americas, and how the European sailors and the Native American inhabitants had two wildly different interpretations of the event. For the European sailors, they were often on the edge of death when they sighted land and realized how close they'd come to disaster. But to the native inhabitants, these men who appeared from the distant horizon were like gods. To Heckenlively, it seemed people were putting the known facts into their already constructed narrative, and throwing away the information which didn't fit that narrative.

> That's exactly right. People are making things up. There's a spectrum of worldviews, political philosophies which manifest themselves from the right to the left, such as authoritarian, reactionary, conservative, moderate, liberal, progressive, and utopian. But what happens is, because people either consciously or unconsciously believe a certain set of fundamental premises about the nature of reality itself, they perceive things differently. And when they get into an area where things are outside of their common understanding, they tend to project speculatively. They project things that are consistent with their other basic assumptions of reality. And the problem is that they then postulate those as though they were true. It's a strange form of existentialism that they don't know it's true, but they act as if it's true.[278]

Sheehan then claimed that many of the early UFO crashes were the result of our radar arrays' interfering with their navigation systems. If true, this raises the inevitable question that if these are intelligent beings, wouldn't they develop countermeasures? Was Sheehan's opinion that the aliens had developed countermeasures to our attempts to bring them down? Heckenlively said that in nature, there's always an arms race. Was there evidence that over the years, the alien ships became less likely to be brought down by our technology? Sheehan's response:

Yes, yes, that's right. You would think that they would have developed countermeasures to us. And that's why the appropriate question is, what is the most recent recovery of these vehicles? And a lot of the accounts of the vehicles that have been taken into possession are quite dated. We're talking about events back in 1945, 1947, and the mid-1950s that some of these vehicles appear to have been recovered. And the response [that] the extraterrestrials may have had was to have at least functionally neutralized that capacity. I don't know. The most recent recovery that I've heard of is really quite old.[279]

Heckenlively next moved to a part of the Betty and Barney Hill abduction story that had left him deeply unsettled. According to their account, the aliens performed invasive procedures on both of them, examining Betty to determine if she was pregnant and harvesting semen from Barney.

In Heckenlively's mind, what they described was an assault, if not an outright rape.

But the Hills didn't perceive it in that manner, apparently because of the calming thoughts the aliens projected into their minds. Sheehan's response was:

You're dealing with a neurolinguistic programming phenomenon, and that's all true. There's no doubt. Any being who can penetrate another's consciousness and communicate something to them effectively has the capacity of seeding into that person's consciousness, ideas or perceptions that are ostensibly foreign to their own nature.[280]

Sheehan went on to say that all five of the extraterrestrial species who appear to be visiting us have faculties of sight, hearing, and touch, as we do, but they also appear to have some sort of telepathic power. He believes that we have this power as well but have not fully developed it. He went on to say that at his newly formed institute, they were trying to avoid presumptions about how benign or bellicose any of these visitors are, and instead doing their best to determine what is true. Of the different species who are allegedly visiting Earth, this is what Sheehan had to say:

We're starting out with a conservative number of at least five different species that have succeeded in coming to our planet.

There are the small grays, who are like three and a half feet tall.

Then there are the tall grays, who seem to be anywhere from five to six feet tall. They appear to be the same species. They look virtually exactly alike, except for their size.

And then there are the reptilians, and these are like five to six feet tall, [with] two arms, two legs, a head, and [a] torso, and they're reptilian.

Next are the insectoids, or mantis people, as they're often called. That's come to be the term often used for them. They're big and tall. Six feet to seven feet tall, kind of stooped over, and very skinny. They remind everybody of a praying mantis. And virtually everybody who has encountered them, often in the company of the small gray beings, or the reptilians, report the insectoids are the ones giving directions. They're in charge.

And finally, there's the wild card, the so-called humans, the Pleiadians, who look like us—tall, blonde-haired, blue-eyed, white people. There's a substantial amount of information about these Pleiadians, but we put a big caveat on it because it just smacks so much of a projection of our own *Homo sapiens*-centered reality that it's hard to believe it's true. I mean, everybody would be in an entirely different state of mind about the UFO phenomenon if they thought they were just people like us.[281]

Heckenlively raised with Sheehan the possibility that the aliens might be afraid of us, that our skills of adaptability might be something rare in the universe. Sheehan agreed that alien fear of us might be a possibility, but took a much more sanguine view of extraterrestrial intentions.

We believe that there have been credible accounts by credible witnesses of the beings from more than one of these species, not just the small or tall grays, or the reptilians or insectoids or humanoids, consistently commenting about the danger of nuclear weapons.

They are, in a sense, trying to warn us that we have to get rid of nuclear weapons because they are a threat to the life-generating capacity of our planet. And perhaps there are a comparatively few life-generating planets in the universe, and they're of great value. And therefore, our presenting a threat to terminate the capacity of our planet to gestate life is something that goes beyond something that is subject to our legitimate, exclusive jurisdiction.

And that is an important postulate of our institute—that there may be a potential legitimate standing [that] extraterrestrial species may have in not allowing us, even as the dominant biological species on our planet, to terminate the life-generating capacity of our planet. That this threat could legitimately justify potential intervention on their part, if necessary.[282]

Heckenlively understood the argument Sheehan was making, but didn't know if he believed it.

Maybe the aliens wanted to protect the Earth and humanity.

Or maybe they had reasons of their own which we don't understand at all, and we should.

🔺 🔺 🔺

As Heckenlively and Sheehan neared the end of their conversation, Heckenlively shifted the topic from the actions of aliens to those of humans. Specifically, Heckenlively asked him, "What is your best guess, after nearly eighty years of trying to reverse engineer this technology, as to how much progress has been made?" Sheehan's response:

I suspect what they've done is figured out some small portion of the technology, and if they power the craft with a small nuclear power generator, they might be able to fly it. I think they've mastered, to some extent, some of the antigravity technology. There are some fairly credible accounts now. But I don't go as far as Dr. Steven Greer goes.

There's one guy, Michael Schratt, who claims we've mastered the technology and can pilot the craft. That they've backengineered the technology and created manmade craft that, as Michael likes to say, "can accelerate like a spark off a grindstone" and go out of sight instantaneously. The speed is so incredible. And they're supposedly manned by US military personnel.

And of course, that's got Greer believing we've already got a program and it's fully developed with multiple craft, and 95 percent of UFOs people see are really man[-]made craft and run by us. And of course, they're going to use them to generate a false flag and pretend we're being invaded by space aliens, and it'll be a financial wet dream for the national security state. It's that kind of thing.

But I do think they've been able to figure out some of the technology and have some craft that can mimic an alien vehicle.

And of course, this opens onto another incredibly important issue. And that is the number of credible accounts of people who, when abducted by either the tall or short grays, say there are US military personnel with them as well.[283]

Great. Now we have to worry about the military and the aliens working together to abduct us, perform experiments on us, harvest our eggs and milk us for our semen, and maybe put our DNA in the grays or the reptilians.

Which group—aliens or the military—should we fear more?

Heckenlively asked Sheehan what he made of such accounts, and he seemed to be as confused and disturbed as Heckenlively felt.

It's a game changer for sure, because that opens onto the question of there being some sort of tacit agreement or treaty. An operating agreement between United States military authorities and representatives of the extraterrestrial species.

The implications of that are so profound that one would have to be careful about that. But I've talked to people I trust who have reported being in the presence of ETs with US military personnel coordinating with them.

And it doesn't take more than one of those examples and all of a sudden, you're in a different world. And so that's an area we're going to be subjecting to some extreme scrutiny.[284]

Could this be the "catastrophic disclosure" that Michael Mazzola had been warned about, which made Heckenlively so interested in coauthoring this book? Distract people with the idea that UFOs are real, and then maybe they won't question the accounts of people who not only were abducted and experimented upon, but who say in addition, that these events often took place in the apparent presence of US military personnel.

Mazzola draws attention to the claim by Jacques Vallée that he was in possession of a CIA document from the 1980s suggesting the CIA was staging fake alien abductions in Brazil and Argentina as part of its psychological warfare operations. Heckenlively was able to find confirmation of Mazzola's claim in the following article:

> Vallée is one of many who have written about and documented the startling evidence that well-constructed hoaxes and media manipulations have misled UFO researchers, diverting them from the UFO phenomenon itself, and what's going on.

> In one of his books, Forbidden Science 4, he shares a record of his private study into unexplained phenomena between 1990 and the end of the millennium, during which he was traveling around the globe pursuing his professional work as a high-technology investor. It's a bit of a diary, documenting his experiences and encounters/meetings as he tries to examine and explore the phenomena.

> In an entry dated Thursday 26 March 1992, Vallée writes: *"I have secured a document confirming that the CIA simulated UFO abductions in Latin America (Brazil and Argentina) as psychological warfare experiments."*[285]

Mazzola claims that in 2021, his production team reached out to Vallée to request the document, and Vallée became hysterical, and refused to produce it.[286] Going directly to Vallée's book for the exact passage:

> I have secured a document confirming that the CIA simulated UFO abductions in Latin America (Brazil and Argentina) as psychological warfare experiments. On Tuesday aerospace researcher Morvan Salez met me in Burbank. I gave him the Fleximage digital tapes of the Costa Rica photographs before

going to dinner in Hollywood with Bob Weiss and Tracy, who told us that the abduction miniseries had run into problems; John Mack [Harvard professor, psychiatrist and Pulitzer prize winner] insists the Aliens are good guys who came to save us, while the plot takes a negative twist in Hopkins [Budd Hopkins was an accomplished artist and later wrote several books about the UFO abduction experience] masochistic mindset. They are on unfriendly terms.[287]

The short passage reveals many of the common patterns one encounters in researching the UFO issue. Claims of government involvement, and individuals of genuine interest in the field, like Mack and Hopkins, drawing radically different conclusions from roughly the same evidence.

It was frustrating that so many questions could be answered only with guesses. But the framework of the problem was coming into better focus. Sometimes the questions seemed to lead to answers that were so bizarre that they mess with the mind, as when Heckenlively asked Sheehan about claims the US was in possession of nine downed UFOs. Here's what Sheehan said.

Yeah, that's right. That's the number that keeps being reiterated by some of the people that are in the whistleblower category, that there's nine of them. And that one or more that they've recovered are operational and not severely damaged. But they've had a terribly difficult time figuring out, number one, how to get into them.

This may be a lot more data than you need, but there appears to have been a significant problem getting into them because they were kind of sealed hermetically. They couldn't find any entrances or anything [for] how to get into these craft, and they were impervious to being penetrated. Except there was one, an incident which occurred—one of them did crash-land, but it was still basically fully intact.

They set up a zone all around it, where they cordoned off the entire area and falsely reported that there was a release of some hazardous material there. And they utilized the authority that they had garnered to themselves to declare martial law around the area as a hazmat crisis. And they brought in a big bulldozer on a big eighteen-wheeler, and they hooked [the craft]... because they couldn't figure out how to get into [it].

And so they were hooking up chains or something to the craft to try and pull it out of the embankment. It had been impaled into the side of the embankment, and they were trying to pull it out.

And when they did, a pie-slice-shaped section of the craft came loose. This was after they'd spent years trying to figure out how to get into the other craft they had.

It came loose from the craft, and they were all panicked and they stopped. They all had to reconnoiter and figure out what the hell to do about all this.... They finally resolved that what they were going to do was send in this guy, a Specialist Fourth-Class guy with a laser arc or lamp, to go inside the craft and see what was inside it.

He went in—crept through this little opening and went inside the craft. The craft was about 35 feet or so in diameter and 15 feet high. But when he got inside the craft, it looked like it was about the size of two football stadiums inside. And he became so totally disoriented because of this reality that he threw up.

And he crawled back out of the craft right away. And what he thought was about two or three minutes was like four hours. So apparently there was not only an extraordinary distortion in space inside there but also some sort of time distortion.[288]

The account sounds like something out of an *Alice in Wonderland* fantasy. But if these aliens have the technology to travel between planetary systems, it shouldn't come as a surprise that many elements of what they possess would appear magical to us.

In Sheehan's telling of the extraterrestrial presence in our world, it would seem that the aliens have a few fail-safes regarding the ability of humans to exploit their technology. After they dealt with how our radar systems interfered with their guidance systems, we had trouble entering their craft and understanding what we saw when we entered.

And finally, according to Sheehan, their craft appeared to be operated through the use of a special helmet placed on the head of the pilot. However, in order to operate the craft, the pilot needs to be at a certain level of consciousness—not necessarily spiritual; let's simply call it an advanced, peaceful mind. According to Sheehan:

> The people that they tried to put the helmet on were not adept enough telepathically to make it work. Anybody who had a low enough level of consciousness to be in the military or intelligence community wasn't in a high enough frequency to run the thing. So, the very people that wanted to use it as a weapons platform were unable to make it work, because they were at too low of a vibrational frequency.

> And they tried to get other people. They eventually got [CIA analyst and remote viewer] Pat Price to do all kinds of things, but Pat Price got murdered. Their top guy got murdered because he was just too dangerous. They pretty much concluded it was the Russians that killed him.[289] [Author's note: according to Sheehan, Price also had been experimenting with locating Russian submarines, and his psychic ability made him a danger to the Russians.]

As Sheehan claimed later in our interview, "There's this almost built-in organic safety device, that you can't operate this technology unless you are of a high enough level of consciousness that you'll use it only for more or less benign purposes."[290]

In January of 2025, insider Jake Barber corroborated and expanded upon Michael Herrera's account. He claimed the individuals Herrera saw being loaded onto the craft were part of a secret program called the "psionics:" psychically gifted individuals, who could telepathically call in these craft and worried about the government's plan for how to use these individuals.[291]

In their interview Heckenlively and Sheehan spent some time discussing a few scary cases of what had allegedly happened to some UFO whistleblowers, and the possibility that Sheehan was disclosing information that might put the authors of this book in physical danger. Sheehan said:

> The reality is that the members of the national security state believe they have been authorized to use force to protect this secret. And they do not believe that they have been confined

in the exercise of that authority to any particular geographical location.

It's not just that they keep them out of Area 51 or S-4, but they've been authorized to use lethal force to protect the secrecy. And the people in the program know that. And so the people are extremely reluctant. It's not just that they're afraid of violating their nondisclosure agreement. They're afraid of being terminated.

And so, this is a very serious problem. People in the program aren't asking, "What is it that needs to be done in order to get this information fully disclosed?" What they're asking is a different question: "How much of this information can be forced to be disclosed, without that element [the national security state] retaliating against the people that are participating in getting the information exposed?" Once again, it's leaving in the driver's seat the members of the national security state, who are the protectors of that secret.

And what we have to do is break through that.[292]

Sheehan's account raises the question: "Who should we fear more? The humans or the aliens?"

CHAPTER NINE

A SKEPTICAL CONGRESSMAN'S VIEW ON UFOS

Heckenlively was at the FreedomFest event in Las Vegas, Nevada, from July 11 to July 14, 2024.

On Friday, July 12, he attended a fundraiser for Michigan US Senate candidate Justin Amash and was fortunate enough to get into a conversation with Congressman Eric Burlison, from the 7th district of Missouri, the "Show Me State," as well as Congressman Warren Davidson, from the 8th district of Ohio.

Like many Americans, Heckenlively had a generally low opinion of Congress members, but these two men changed my mind. In their telling, when federal employees are called to testify or produce documents for a congressional inquiry, they often seem to consider themselves, rather than Congress, the actual government. The congressmen told him they were essentially powerless to punish federal employees who failed to provide requested documentation or gave evasive replies to their questions. The behavior of the federal employees they saw rarely rose to the level of open defiance; rather, it was a siege in which their delaying tactics and general lack of response would hopefully exhaust the Congress members, and they would move on to the next issue.

When Congressman Burlison mentioned the difficulty he was having as a member of the Oversight Committee, particularly when dealing with the UFO issue, Heckenlively's attention immediately perked up.

Heckenlively told Burlison he was writing a book on the subject of UFOs and asked if he would consent to a recorded interview. He agreed, and Heckenlively was delighted.

The following is an account of that interview.

⋏　⋏　⋏

Here's a little background information on the congressman, in his words.

> I was a computer nerd throughout high school and college and paid my way through college building websites. I then started writing code for a hospital system in the area, working as a consultant for Oracle and Cerner [a US-based multinational provider of health information technology], then Oracle itself, and along the way found I loved finance. I was always the guy at the water cooler who would be asked for what investments to make. I was doing my own day trading and leveraging a lot of what I'd learned in my college finance classes. Eventually, I realized I needed to switch careers.

> [On my own,] I took and passed the Series 65, and became a registered investment advisor, started taking clients, and began building up a portfolio that was pretty robust. Once word got out that there was an advisor who actually manages options and does option strategies, I started becoming popular. So, I did that for a while.

> I've always been involved in politics, then became a state legislator in Missouri for eight years. Then I was a state senator for four years. When the congressional seat in my district opened up, I ran for it and won. That was in 2022.[293]

Prior to becoming a congressman, Burlison had spent twelve years in state government. At the time we were speaking, he was coming close to finishing his first term in Congress. I asked him to tell me about the House Oversight Committee and its duties.

> Oversight is the investigatory arm of Congress. And that's anything. In the 1990s, when there was the issue of steroid usage in baseball, Oversight is the committee that had baseball stars like Mark McGwire, Sammy Sosa, and all of these other players brought before Congress and asked about their steroid usage.

> When I first got elected [and placed on the Oversight committee], my first hearing was about Twittergate. We subpoenaed

and brought forward all these executives from Twitter. Some of them had quit already. Some of them were about to be fired by Elon Musk.

But we were able to get them on the record telling us how they worked with the federal government to suppress Hunter Biden's laptop, to suppress and block conservative accounts, and all of that.[294]

While Burlison is supportive of the mission of the Oversight Committee, he wondered if its members have the competence and background to properly investigate the UFO issue, considering that there have been accusations of intelligence agency involvement and the collusion of major aerospace corporations.

Most of what the committee does is investigate. For example, we were investigating the Biden family and all of their financial transactions. A lot of the staff on Oversight are not geared up for a UAP discussion. They're not from the intelligence community. I'm trying to be respectful of them. They're experts in looking for fraud. They're not experts in tracking down a top-secret government agency and the documents that agency might have.[295]

The congressman later expressed his opinion that the investigation would be much more authoritative if it had individuals with intelligence or military backgrounds. But they don't, and that is a drawback.

Heckenlively asked him about David Grusch, the whistleblower who'd first piqued his interest in this subject, as Grusch seems to have the credentials that would, even to the most skeptical of individuals, give an air of plausibility to the accounts of crashed alien ships and government-sponsored reverse engineering programs. His response:

I saw him on the news and told my staff, "Let's get him in here." We were all watching it together. Because on the news he said, "I'm doing this so that someone in Congress would see it and be able to hear me." We reached out and had a call with him.

I thought he was credible.

I and other members of the committee referred it to the chairman, Jamie [James] Comer. He agreed and decided to have a hearing on it. But he didn't want to use the whole Oversight committee.

We used one of the subcommittees to investigate, chaired by Glenn Grothman out of Wisconsin. Tim Burchett was kind of the lead on organizing and was there, but at the end of the day it was Glenn Grothman.

The hearing was pretty interesting. I come from the "Show Me State." I'm very skeptical. You're going to have to show me before I believe all this stuff, okay? I want to see it.

Now, at the hearing [was] David Grusch, along with pilots who had seen things and told us what they saw. I believe these guys saw something on the radar and visually, what they saw was real. My question is, just because you see some object and it looks strange, does that necessarily mean it comes from another planet?

That's a conclusion I don't think we should jump to until we have all the facts.

What I think is more probable is that what we're seeing is coming from our intelligence agencies, our military-industrial complex, or adversaries, like China or Russia. And I said that at the committee hearing. I think I was probably the one skeptic in the room that day.

But I think people within the UAP community respect my position, because they know I'm not gullible. And honestly, if I ever come out and say, "There are aliens," you will know that I absolutely believe it, and that something's changed. That's the position I have. I think it gives me a credibility in this arena, being skeptical.

However, at the end of the day, what Grusch said deeply disturbed me. The fact that our pilots are having these near misses and we don't know that they are—that alone should be cause for an investigation.

In addition, when you hear reports from Grusch or the other whistleblowers that there are these Special Access Programs that Grusch was investigating, and they're not revealing information to Congress but they're spending our money, it concerns me. There are completely black programs, and that's not okay in my book.

We investigated that claim. We had a briefing with the Office of Inspector General for the Department of Defense, as well as the Office of the Inspector General for the Intelligence Community.

And without saying anything that's top secret, we did confirm David Grusch is correct when he says there are programs hiding information from Congress and other agencies. David Grusch is correct on these things.[296]

Heckenlively brought up the suspicions voiced by others that these UFO hearings were creating an "alien threat" narrative that would be used by the military-industrial complex to justify additional spending. Burlison's response:

I suspect that this could be a psyop—that it is an effort by the Department of Defense and the intelligence community to get more money. I am cognizant of that and aware and try to be cautious about that. I think if we're spending money on these programs, then these programs should be accountable to Congress. They should be giving us briefings. We shouldn't have to go searching and ask the right questions, right?[297]

Heckenlively asked Burlison to clarify what he meant by having to "ask the right questions." His response:

The hard part is, I don't even know what to ask sometimes. You don't know until you have a little bit of information and you can further probe. But they're not going to give it to you. They're not going to just walk in and say, "We understand what you want, so here's the briefing." You have to ask the right questions.[298]

Heckenlively was finding this resistance from the federal agencies called before Congress to be perplexing. Heckenlively told Burlison it seemed a congressman should simply say, "Give me everything you've got on UFOs since 1945," and the agencies would comply. Burlison shook his head and continued to explain.

> For example, we would have to say, "Well, what about the event on the [*Nimitz*] on this date. What was that?" Or, "tell us about the Tic Tacs specifically," and on this date and these sightings. So you have to get into the details.

> We had a question of, "Are there sites that are storing top-secret information about these? [Author's note: meaning the Tic Tac–type craft spotted by pilots and personnel from the USS *Nimitz*.] That's what's in Grusch's report.

> We were given locations by one Office of the Inspector General, who said that in his investigations of those sites he didn't find anything. But there are documents and other things in these sites. Those sites are listed.

> But it was made very clear to us that the names of those sites are top-secret. I don't even know how to go back to my office and take the next steps to investigate.[299]

What followed next was an almost comical exchange, as the two men danced around what may or may not be top-secret information without revealing such information.

As best as Heckenlively understood what Congressman Burlison told him, it went something like this:

First, he knows the names of top-secret bases where there is certainly information about advanced aeronautics programs and maybe material related to UFOs.

Second, at least one inspector general's office has investigated these sites but has not found any useful information or material.

Third, the congressional members who know the names of these secret bases are sworn to secrecy as to their locations and names, and would probably be breaking their security oaths if they reveal the names or locations of these bases, even in the context of a congressional investigation.

In other words, Congress has some information but cannot use that information as the basis for any further investigation.

Heckenlively asked Burlison his opinion about how the intelligence community treats congressional members, and he had a lot to say.

> I think that generally speaking, we don't get good briefings. Often, I walk out of the briefing, get my phone back, look at the news, and find out what I was just briefed on is all over the news. I'm not given access to information that hasn't been leaked in some way. And so what's disturbing is, there's been times I've gone to a briefing, not on UAPs but on something the leadership or intelligence chairman wanted people to be briefed on. And to get the briefing, I have to sign a document saying I won't disclose the information under penalty of perjury.

> And then when I walk out of the briefing, it's all over the news. But ironically, since I got the briefing, I can't talk about it, even though it's all over the news. The news can talk about it, but I can't.[300]

In Burlison's view, Congress members often are trying to do the right thing but find themselves stymied by the intelligence community or other agencies. It was quite something to hear a sitting member of Congress reveal such bitterness only a year and a half into his first term. Given such a bureaucracy that seemed averse to change, or even transparency, it was easy to understand how wild theories could quickly develop.

Heckenlively shifted the interview to claims that private-sector companies might have created these craft that are being seen by the military, such as the claim that Lockheed-Martin might have developed the Tic Tacs seen by the crew of the USS *Nimitz* off the coast of San Diego, California, in 2014. Heckenlively suggested that if the claim is true, it means our military contractors were interfering with the operation and safety of our armed forces. Congressman Burlison said:

> I agree. I totally agree. I think that it may be that the private sector is testing the capabilities of these devices, and whether they can be seen by our military. They may be doing that. But I don't know. What I can say is that some of these things that have been seen are low-tech.

For example, what's very well-known publicly are these Chinese spy balloons that floated across the United States. What I can say from these meetings is that China is much more creative and aggressive than anyone could have ever guessed.

And then, the other thing is that it's not just high-altitude balloons. They've got other creative ways of getting drones and other things up, that can be spotted. They're sending them to our military bases, and that's a concern. That concern, in my opinion, is something that should be investigated.[301]

As Heckenlively came to the close of the interview with Congressman Burlison, he wanted to make sure he understood the specific points the congressman had told him. Regarding the AARO report, it seemed that whistleblowers were credible individuals, they had seen technology from hidden programs, but the government claims they had misidentified our advanced technology as being from aliens. The congressman replied:

Yes. And that was one thing they did give us specifics about. And I can understand why they were mistaken in what they believed.[302]

If this was true, it means that the intelligence agencies were allowing these people to falsely testify in public, prior to revealing to members of Congress that the information was mistaken. In addition, it means that the intelligence agencies would be engaged in letting a little information out about these secret programs but keeping a great deal of it hidden.

Heckenlively found Congressman Burlison to be an objective evaluator of the information and a person who sought to find common ground with the UFO community, as when he acknowledged that a certain number of reports remain unexplained.

I learned something that confirms my worldview, which is that we were able to debunk and demystify countless numbers of events and sightings. Now, not all of them. But the vast majority. And the truth is, when I talk to believers in UFOs who are in the UAP community, the people with any kind of credibility will say the same thing. They'll say, "The vast majority of the people, the photos, or whatnot, are not a real visit from an alien. It's not really a UFO."

And that makes sense. But I still have yet to have the experience where I see evidence for something that's not explainable.[303]

Heckenlively considered that to be a fair point. Burlison hadn't seen any evidence that has convinced him beyond a reasonable doubt that UFOs or aliens are real.

One of the most fascinating areas of agreement between the two was about the unconstitutional power that the intelligence agencies seem to be asserting over our government. The authors believe this is one of the most critical aspects of the story.

> **KENT HECKENLIVELY:** So, it's your current understanding that the intelligence agencies, and by extension certain private industries, are concealing information from Congress? But you don't know what that information consists of?

> **CONGRESSMAN BURLISON:** Correct.[304]

Determining the importance of the information behind this wall of secrecy is difficult, because there are conflicting interpretations—some benign, and others sinister. Indeed, the entire UFO issue can be considered something of a psychological Rorschach test, in which people see what they want to see in a series of inkblot images, or unexplained reports.

On the one hand, our military-industrial complex has a valid and defensible interest in keeping its secret programs safe from discovery from our potential adversaries. It's reasonable to conclude that keeping members of Congress in the dark about such programs is an understandable approach, given that information from Congress is leaked so easily, as we see in our daily news.

But on the other hand, if someone wants to keep alien technology hidden, either to create a program like Prompt Global Strike—which, as described by Sheehan, would allow a nuclear weapon to be delivered to China or Russia in a few minutes—or Project Golden Dome, which identifies alien vehicles operating in our atmosphere, claims of secret human programs would be the perfect place to conceal such information.

And it appears there is not a clear political divide among those who believe we should and those who believe we shouldn't be getting this information. Some Republicans, such as Congressman Burlison, are interested in disclosure, while other Republicans oppose the effort. The same can be said of the Democrats. This exchange occurred near the end of the interview:

> **CONGRESSMAN BURLISON:** Honestly, I'll give the Democrats some credit. This is kind of a bipartisan effort. We've got Democrats that are on board with trying to get to the bottom of this.

> **KENT HECKENLIVELY:** Yeah, I understand [Senate Majority Leader] Chuck Shumer is a big proponent of this effort.

> **CONGRESSMAN BURLISON:** Yeah, in honor of [former Majority Leader] Harry Reid. He's a proponent of it, trying to get to the bottom of the information. What's disturbing is that when we tried to get language in the Department of Defense bill, we had the intelligence community, in the form of the chairman of the Intelligence Committee, Mike Turner, who tried to do everything he could to stop us from getting access to the information.[305]

Does it genuinely matter whether one is a skeptic or a believer in UFOs? What seems clear to both sides is that something important is being kept from us.

The question we must ask ourselves is whether the public is best served by continuing to keep that information hidden, or by knowing the truth and discussing it as reasonable adults.

The authors believe you can handle the truth, whatever that might be.

CHAPTER TEN

A CONGRESSIONAL HEARING TURNS INTO THE THEATER OF THE ABSURD

On November 13, 2024, Chairwoman Nancy Mace convened a joint hearing of the US House Oversight Subcommittee on Cybersecurity, Information Technology, and Government Innovation, as well as of the Subcommittee on National Security, the Border, and Foreign Affairs, to discuss new information related to the UFO/UAP issue.

A whistleblower had come forward with claims of a government program called Immaculate Constellation that possessed high-quality photos, videos, and sensor data of unidentified craft. The images were supposed to be much crisper, sharper, and more detailed than anything previously available to the public. After making her introductory remarks, Congresswoman Mace opened with an astonishing statement.

> Former National Security Advisor H. R. McMaster said on Bill Maher's program that "there are phenomena that have been witnessed by multiple people that are just inexplicable by the science available to us." Army Colonel Karl Nell, a member of the Federal Government UAP Task Force, said at a conference this past May that nonhuman intelligence exists; nonhuman intelligence has been interacting with humanity. This interaction is not new, and it's been ongoing. And there are unelected people in the government that are aware of that.

> But UAPs remain a controversial topic. I'm not going to name names, but there are certain individuals who didn't want this hearing to happen because they feared what might be disclosed. But we stood firm. No amount of outside pressure would ever

keep me from pursuing this subject to ground, come hell or high water.[306]

Even a UFO skeptic would likely have a difficult time understanding why individuals in the government wouldn't want to discuss this issue. The government either has convincing evidence of UFOs or it does not.

Several Congress members gave opening statements that revisited alarming sightings, but probably the most disturbing was the account by Congressman Glenn Grothman:

> I'm deeply alarmed by the reporting of the massive drone swarm that flew over Langley Air Force Base in Virginia last December. Langley is the home of the First Fighter Wing, which maintains half of the F-22s in the US Air Force inventory. Reports of this incident indicate the drones were roughly twenty feet long, flying more than a hundred miles an hour at an altitude of three thousand feet, yet the origin of these drones and their operators remains a mystery.

> The incident and other sightings near sensitive military installations highlights the complexity of the UAP challenge facing our Intelligence, Defense, and Homeland Security committees. Whether these phenomena are the result of foreign adversaries developing advanced technologies or something else entirely, we must take them seriously, investigate them thoroughly, and assess their implications on national defense.[307]

When Heckenlively read the testimony, he was of a divided mind. One part of his brain was thinking it was remarkable that these events were being discussed so publicly. But another part kept thinking of the admonition of Steven Greer that this supposed disclosure was designed by the intelligence agencies to generate a threat narrative to feed the military-industrial complex.

Why were we being given only two options: a foreign power or an alien one?

Shouldn't it at least be on the table that perhaps our own defense contractors are spoofing our military to create pressure in Congress for additional funding of their pet projects?

The opening statement of Congressman Jared Moskowitz of Florida accurately captured part of his frustration.

> When the American people and members of Congress ask, "Are reports of UAPs credible," we're met with stonewalling. We're met with responses of, "I can't tell you." And in fact, we're met with people not wanting us to have hearings. We're met with people not wanting us to ask you questions. In fact, many of us, we're told not to ask some of you certain questions on certain topics.
>
> In a time of heightened distrust of our government institutions, I believe more transparency is not only needed but is possible. And obviously we can respect national security limits, but we also have to provide our constituents with the information and oversight that they have tasked us for.
>
> What are UAPs? Are they real? Are they ours? How has this technology been funded? How do they get funded?[308]

If one has seen a UFO, there's no doubt in that person's mind the phenomena is real. But the majority of the American public has not seen a UFO. How are they to assess the information?

What should be troubling to all are questions that cannot be asked or information that is unreasonably withheld.

To the credit of the Congress members holding the hearing, they appeared to be trying to get to the bottom of the issue by bringing before them the most credible individuals possible. The first of the four witnesses to testify was Rear Admiral Timothy Gallaudet, US Navy (retired). He said:

> Confirmation that UAPs are real came to me in January of 2015 when I was serving as the Commander of the Navy Meteorology and Oceanography Command. At the time, my personnel were participating in a predeployment naval exercise off the US East Coast. It included the USS *Theodore Roosevelt* Carrier Strike Group, and this exercise was overseen by the United States Fleet Forces Command led by a four-star admiral, who at the time was my superior officer. During this exercise, I received an email in Navy Secure Network from the operations officer

of US Fleet Forces Command. The email was addressed to all the subordinate commanders and the subject line read in all capital letters, "Urgent Safety of Flight Issue."

The text of the email was brief but alarming, with words to the effect, "If any of you know what these are, tell me ASAP. We are having multiple near midair collisions, and if we do not resolve this soon, we are going to have to shut down the exercise."

Attached to the email is what is now known as the "GoFast video" captured on the forward-looking infrared sensor of one of the Navy FA-18 aircraft participating in the exercise. The now-declassified video showed an unidentified object exhibiting flight and structural characteristics unlike anything in our arsenal.

The implication of the email was clear. The author was asking whether any of the recipients were aware of classified technology demonstrations that could explain these objects. Because the DoD policy is to rigorously de-conflict such demonstrations with live exercises, I was confident this was not the case. The very next day, that email disappeared from my account and those of the other recipients without explanation….

I concluded that the UAP information must be classified within a Special Access Program managed by an intelligence agency that is a compartment program, that even senior officials, including myself, were not read into.[309]

Time and again we keep coming back to a familiar narrative: top military and governmental officials are having information withheld from them without explanation.

Leading officials in our military are interacting with these craft, and they are baffled as to what they might be. Attempts to gather further information are met with a stone wall of silence.

The next to give his statement was Luis Elizondo, the former head of AARO and author of the 2024 book *Imminent: Inside the Pentagon's Hunt for UFOs.*

Let me be clear, UAPs are real. Advanced technologies not made by our government or any other government are monitoring sensitive military installations around the globe. Furthermore, the US is in possession of UAP technologies, as are some of our adversaries. I believe we are in the middle of a multidecade secretive arms race, one funded by misallocated taxpayer dollars and hidden from our elected representatives and oversight bodies. For many years, I was entrusted with protecting some of the nation's most sensitive programs. In my last position, I managed a Special Access Program on behalf of the White House and the National Security Council. As such, I appreciate the need to protect certain sensitive intelligence and military information. I consider my oath to protect secrets as sacred, and I will always put the safety of the American people first. With that said, I also understand the consequences of excessive secrecy and stove piping....

A small cadre within our own government involved in the UAP topic has created a culture of suppression and intimidation that I've personally been victim to, along with many of my former colleagues. This includes unwarranted criminal investigations, harassment, and efforts to destroy one's credibility. Most Americans would be shocked to learn that the Pentagon's very own public affairs office openly employs a professional psychological operations officer as the singular point of contact for any UAP-related inquiries from citizens and the media.[310]

Despite the misgivings of many, Elizondo could be an authentic whistleblower, and if so, he has the background knowledge to give great weight to his claims. Has the UAP program become so riddled with inconsistencies and hypocrisy that anyone with courage and a conscience is driven by the need to become a whistleblower?

A remarkable and somewhat frustrating series of questions by Congressman Eric Burlison encapsulates the dilemma of what to believe. Congressman Burlison was trying to get to the bottom of some material from an alleged crashed UFO, which supposedly collected in the 1950s. It was unclear whether this material was with Lockheed-Martin, Bigelow Aerospace, or the CIA.

CONGRESSMAN BURLISON: So, if this material exists today, who's in possession?

LUIS ELIZONDO: Sir, I wouldn't be able to have that conversation in an open hearing. We'd probably have to have that [inaudible].

CONGRESSMAN BURLISON: Okay. My question to you then is, if we were in a secure setting, if we were in a SCIF [sensitive compartmented information facility], would you be able to provide, or get access to something, whether it's visuals or material that we could put our hands on, or biologics, that would convince me, that would show to me they were of nonhuman origin?

LUIS ELIZONDO: Sir, that decision would not be mine. That would be the gatekeepers still in the US government.

CONGRESSMAN BURLISON: So, if you were in our shoes, where would you go from here? Lots of times we just don't know who to ask because we don't know where to go next. If you were in our shoes, where would you go?

LUIS ELIZONDO: Well, I prefer to answer that question in a closed session. However, we established AARO for that very purpose. And unfortunately, under its previous leadership it failed. So, one would hope that they would have the authority necessary to do that. Let's hope this new iteration of leadership will be successful.

CONGRESSMAN BURLISON: It was previously testified that there were biologics that were collected. Are you aware of any of that?

LUIS ELIZONDO: I am, sir, aware of the reporting that biologics have been recovered. Again, my focus was more nuts and bolts—looking at the physical aspects of these phenomenon, how they interacted around military equities and nuclear equities. So, I'm certainly not a medical expert. I would not be able to properly provide you with a whole lot of value in that. Simply because I don't have the expertise.

CONGRESSMAN BURLISON: Was anything described as that "we have possession of bodies"?

LUIS ELIZONDO: Yes. Yes.

CONGRESSMAN BURLISON: Is it multiple types of creatures or...?

LUIS ELIZONDO: Sir, I couldn't answer that. I can tell you anecdotally that it was discussed quite a bit when I was at the Pentagon. The problem is that the supposed collection of these biological samples occurred before my time. In fact, before I was born.[311]

There's a lot that's concerning in this exchange between a congressman, charged by the country's founding documents with conducting oversight of money spent by the government, and an employee of that government, who has sworn a secrecy oath to the very same government.

Again, one is reminded of Greer's idea that this secrecy is a direct violation of our Constitution.

The civilian branch of our government is supreme over the military and intelligence agencies, or else we do not have a representative government. The above interaction suggests that we may have only the facade of a democratic system, and the illusion remains only as long as no one points out the glaring inconsistencies.

In the answers given by Elizondo seems to be three categories of questions.

First, there are questions Elizondo can answer in an open congressional hearing.

Second, there are questions Elizondo can answer only in the privacy of a SCIF.

And third, there are questions that Elizondo cannot answer in a SCIF; rather, he would have to contact some unidentified "gatekeeper" in the government who either would give permission to answer or would be able to provide the information.

For the typical citizen used to our system of justice in which a witness will raise their hand and promise to "tell the truth, the whole truth, and nothing but the truth," this division of questions that can and cannot be answered in a public setting seems puzzling.

And this curious bifurcation of our government into sections that can and cannot be talked about in public seems to almost take a back seat to the startling answer given by Elizondo that these anonymous individuals in the government talk quite openly about the recovery of "nonhuman biologics" and that the craft in which they travel is regularly interacting with our "military equities and nuclear equities."

Further questioning of Elizondo by Congressman William Timmons of South Carolina dealt with the question of how we should respond to these incursions.

CONGRESSMAN TIMMONS: Mr. Elizondo, you just said something interesting. You said they don't seem to be hiding. The UAP sightings are becoming increasingly brash, if you will. And we've been hearing about this for years. But they've generally been isolated, and not as consistent, and over critical military installations. Would you say that's fair? Is it happening more and more?

LUIS ELIZONDO: Great question, sir. Let me see if I can answer this for you. Certainly there seems to be some indication that they're being provocative. Meaning that they're in some cases literally splitting aircraft formations right down the middle. So, that's an air safety issue.

The question is, is the frequency increasing? And really, the response is, it depends. Yes, it's possible that there's an increase in frequency. But it's also possible that there's heightened awareness now. And there's also more pervasiveness of technology out there that's collecting this information and can record the information.

So, we're not quite sure yet if it's actually an increase in numbers of these events, or is it that we have better equipment now to record these things and we have a better ability, if you will, to analyze these things?

CONGRESSMAN TIMMONS: And that's my next question. It seems that a lot of these sightings occur near military installations. Do you think these UAPs are intentionally targeting military installations, or do you think that we have increased abilities to monitor [areas] surrounding the military installations?

LUIS ELIZONDO: Sir, maybe both. Part of my concern is we have something in the Department of Defense, in the intelligence community, called IPB, initial preparations of the battle space. And we use equities like ISR (intelligence, surveillance, and reconnaissance), and other types of technologies to prep the battle space. And certainly, if I was wearing my national security hat, even if there was a 2 percent chance that there was some sort of hostile intent here, that's 2 percent higher than we can really accept. And so we must figure it out. There's a calculus, capabilities versus intent, in order to identify if something is a national security threat. We've seen some of the capabilities, yet we have no idea of the intent. And so, this is why this discussion is somewhat problematic from a governmental perspective. Because we have no idea.[312]

Much of what Elizondo said in this section of his testimony strikes one as reasonable. If one sees remarkable capabilities, then one must naturally raise the question of intent. Any other response by those managing the security and defense of a country would be irresponsible.

However, what's missing from Elizondo's account is an acknowledgment that however long aliens may have been visiting Earth, there's no clear-cut example of hostile intent. Given what we know, it seems unreasonable to believe we're facing any imminent or long-term threat of an alien invasion or takeover. Nonetheless, what about those abductions, and the often-reported taking of biological specimens from men and women, including sperm and eggs?

One can't fault Elizondo for his concern that even if there's a 2 percent chance that these entities are hostile, we need to take immediate action.

It's possible that the men in the shadows are the good guys, or at the very least, doing what they believe to be in the best interest of humanity.

⋏　⋏　⋏

The third witness for the hearing was independent journalist Michael Shellenberger. He had seen and reported on what was purported to be a document written by a government UAP whistleblower, pertaining to the alleged secret program called Immaculate Constellation mentioned earlier.[313] The document begins:

IMMACULATE CONSTELLATION is an Unacknowledged Special Access Program (USAP) established following the public disclosure of the AATIP/AAWSAP programs by Luis Elizondo in 2017. Upon disclosure to Congress, it was determined that this USAP and its collateral information have not lawfully been reported to Congress.

IMMACULATE CONSTELLATION's primary mission is collecting imagery intelligence on Unidentified Aerial Phenomena (UAP) and ARV/RV (Reproduction Vehicles) utilizing tasked and untasked US military-intelligence resources. As part of a network of SAPs linked to Non-Human Intelligence (NHI) and UAP technologies, IMMACULATE CONSTELLATION acts as a nexus for collecting, analyzing, and disseminating intelligence on the activities, capabilities, and locations of anomalous aerospace threats that originate from foreign or unidentified sources.

The intelligence within the IMMACULATE CONSTELLATION program primarily consists of high-quality Imagery Intelligence, (IMINT) and collateral Measurement and Signatures Intelligence (MASINT) of UAPs and ARV/RVs within Earth's atmosphere. The collection platforms involved are a blend of tasked and untasked capabilities in Low Earth Orbit (LEO), the upper atmosphere, military and civil aviation altitudes and maritime environments. IMMACULATE CONSTELLATION pays particular attention to anomalous aerospace platforms that have been developed through the study or acquisition of technologies of unknown origin by foreign nations or unknown entities. UAP and ARV/RVs are operating around the globe, often in close proximity to sensitive foreign assets and locations.[314]

There's a lot to digest in those three opening paragraphs, and although the allegations are explosive, one can't help but wonder if the fact that we're writing about them signals that we've fallen into an elaborate intelligence agency trap. However, the fact that the authors downloaded this document from the website of Congresswoman Nancy Mace means that at least this lie has been perpetrated on Congress as well.

If we believe the document to be real, the implications are profound.

First of all, the document claims that UFO information has not been shared with Congress and the American people, a direct violation of existing law.

Second, not only are there UFOs (UAPs), but there are known alien reproduction vehicles (ARV/RVs) operated by the United States or foreign powers.

Third, there are a number of Special Access Programs (SAPs) that are investigating the question of nonhuman intelligence (NHI) and unidentified aerial phenomena (UAP).

Fourth, the Immaculate Constellation program seems to be the umbrella organization for all other Special Access Programs.

Fifth, our most powerful surveillance tools, such as imagery and sensors, have allowed us to identify many of the characteristics of these craft, as well as their preferred operating zones.

Sixth, in addition to nonhuman craft, there appear to be a number of alien reproduction vehicles, designed by humans, that are operating in the same domain.

And finally, these presumed alien craft, as well as the ones humans have reverse engineered, are interested in both foreign and US military sites.

This is the concluding paragraph of the summary section of the report:

> In conclusion, IMMACULATE CONSTELLATION shows that the USG is not only aware of UAPs and TUO [technology of unknown origin], but also foreign state efforts to replicate UAP and TUO capabilities. The data within IMMACULATE CONSTELLATION reveals the capacity of the U.S. Armed Services and Military Intelligence Community to detect, track, identify, and engage anomalous trans-medium platforms. IMMACULATE CONSTELLATION also demonstrates the extant capability to detect, quarantine, and transfer UAP and RV collection incidents before they are observed and circulated within Military Intelligence Enterprise, partially explaining why many otherwise cleared members of the military and IC are unaware of UAP activities. Finally, the existence of IMMACULATE CONSTELLATION provides verifiable evidence of the witting participation by elements of the U.S.

Armed Services, Defense Civil Services, and the Intelligence Community in a global surveillance and reconnaissance mission tasked with monitoring UAPs and ARV/RVs.[315]

One of the challenges when investigating allegations involving the intelligence community is to determine whether there's an inherent internal logic to the claims. If there are craft of unknown origin in our skies, it makes sense that our military and intelligence agencies would develop a program to gather more information about them.

By the same token, if the military and other entities have developed advanced technologies and do not want to disclose this information to potential adversaries, it makes sense that they would come up with a cover story to explain such sightings.

The following paragraph on the imagery of these craft explains Shellenberger's trust in the material.

The IMINT [Imagery Intelligence] collected from datasets available to the DoD, and reviewed for this report, provide compelling evidence for UAP which defy prosaic explanations. There is a large number of unique imagery sensors available to the U.S. military and intelligence community including: Infrared (IR)/Forward-Looking Infrared (FLIR), Full Motion Video (FMV), Thermal, and Still Photography. The multitude of wavelengths collected by the sensors have captured UAP characteristics that are difficult or impossible to observe with the human eye alone. Subtle atmospheric effects associated with UAPs are visible through the sensors employed by the U.S. military and intelligence agencies, enabling unique analytic techniques. The verifiable chain of custody for UAP IMINT collected by U.S. military assets ensures a high level of confidence in the accuracy and integrity of the data gathered.[316]

At the very least, this establishes that the photos and videos collected, as well as the sensor data, have not been tampered with by outside entities. (Although the question remains whether the intelligence agencies themselves may have tampered with the data.)

The document then proceeds to list nine instances in which unidentified objects have been captured by the Immaculate Constellation array of sensors. What's intriguing is that in two of the nine instances, analysts identified the craft as alien reproduction vehicles, meaning they were reverse engineered and presumably operated by humans. This is a sampling:

> **CENTCOM Cuboid Formation of Metallic Orbs**: On USG networks, there exists daytime FMV and daytime-FLIR footage of approximately 12 metallic orbs skimming the ocean's surface at high speed before dispersing in multiple directions. The rapid and agile maneuvering of the metallic orbs were incompatible with known aerospace vehicles and were between 3-6 meters in diameter. In the opening segment of this footage, the approximately 12 metallic orbs flew in a tight "cuboid" formation; the metallic orbs were in three vertical-square formations of approximately 4 orbs each, arranged in a three-pronged configuration, creating the illusion of a cube shape at a distance. All the orbs were white-hot against the black-cold ocean in the FLIR footage, and each sphere created a faint atmospheric distortion both around itself and as a heat-shimmer "contrail." The metallic orbs moved in this cube formation for some time, before rapidly breaking formation as pairs. The sensor platform lost track of most of the metallic orbs as they ascended in altitude and accelerated in speed but maintained observation on a pair of metallic orbs continuing the original trajectory of the larger formation.[317]

The orbs are the greatest mystery to us. We can't imagine their being occupied by any pilot, and the best explanation we can provide is that they're some sort of advanced drone technology. We recall the "foo fighters" that seemed to surround our aircraft in World War II as possible early examples of humanity's becoming aware of these craft. Perhaps now, as humanity is experimenting with drone swarms being utilized by our own drone operators, this phenomenon will become more understandable.

One is much more comfortable dealing with human deceit and trickery, which is why the following account of a possible alien reproduction vehicle is significantly easier for us to comprehend.

> **INDOPACOM Intelligence Vehicles Positioned to Collect on Reproduction Vehicle:** On USG networks, there exists infra-

red footage and imagery of a grouping of vessels engaged in SIGINT and MASINT collection at night in a specific area of the Pacific Ocean. In this footage, which was in close-proximity to the vessels, a large equilateral-triangle UAP suddenly appears directly over the ships. Three bright points are seen at each bottom corner of the UAP, which is observed to slowly rotate on its horizontal access. This rotation partially reveals a horizontal bar of sweeping lights. Intelligence analysis associated with this event specified that the equilateral-triangle is a Reproduction Vehicle (RV) and concludes that the vessels must have been aware of the RV's frequent use of those coordinates, due to foreign pre-positioning of advanced collection assets at the exact time and place. After a brief period of hovering and slowly rotating approximately 500-1000 meters above the ocean, the RV suddenly disappears, and the footage ends.[318]

Here's how we interpret the previous passage: We have secret technology. We wanted to test that secret technology. We sent our military vessels to a specific area of the ocean and waited for the appropriate time to send our secret technology to that location to test how well we can detect our secret technology. Our adversaries also have this information, and so they sent their vessels to a nearby location to monitor our testing of our secret technology.

The military knows about our secret technology, as do our adversaries. But the citizens of the world just get fed lies that the unidentified objects that appear over military maneuvers are a complete mystery.

The next report makes us want to spend however many trillions of dollars would be necessary to make sure we have an adequate defense against the aliens in the event that things fall apart.

INDOPACOM *Large Disc Using Clouds as Concealment:* On USG networks there exists OPIR [overhead persistent infrared] footage of a large saucer shaped UAP emerging from within a dense cloud formation. The saucer registered black-hot against white-cold, with atmospheric disturbances caused by the saucer shaped UAP visible. The saucer was between 200-400 meters in circumference and displayed symmetrical concavities on the upper surface. The saucer shaped UAP emerges at a shallow angle traveling upwards towards the upper atmosphere. After breaking above the cloud cover, the saucer shaped UAP suddenly reverses its direction, descending partially back into the cloud cover, then accelerating rapidly out of frame and partially obscured by the cloud tops. This behavior was evasive in nature and implied that the saucer shaped UAP had become aware that it was under observation by a space-based collection platform.[319]

A flying saucer between six hundred and twelve hundred feet in circumference? This is truly the stuff of nightmare science fiction movies. If the aliens' intentions are benign, what's with the hiding? It seems to us that the claim that aliens conceal themselves in order to not interfere with our development is undercut by the fact that they often seem to do such a bad job of it. If their technology is as advanced as many claim, they should be able to keep themselves completely veiled from us.

The simple fact is that we do not know the aliens' intentions.

Perhaps the aliens hide from us because it is in their interest to do so. Humans' superior numbers and primitive weapons might be enough to overcome their relatively small numbers and advanced weaponry.

Near the end of the report, the author lays out what they believe to be true about the actions of other countries:

- Foreign countries are known to have observed UAPs whose signatures and behavior correlate to those observed by the United States.

- These UAP events are treated by the security apparatuses of each state as serious national security threats due to UAP in proximity to sensitive military and intelligence facilities.

- These facilities are most often associated with aerospace defense, strategic deterrence, and military-sponsored scientific research and development.

- On multiple occasions, each of these nations have attempted to intercept and shoot down UAPs violating their territorial airspace, and the airspace over sensitive facilities.

251

- Foreign countries have internal organizations dedicated to studying the ambiguous threat posed by UAPs, deducing scientific principles through observing UAP, and the careful management of public perceptions of the UAP issue.[320]

It can be extremely frustrating when the people who've been lying to the public for decades decide to tell the truth. I guess that in order to figure out whether what's being said now is true, you have to fall back on what you know about people and organizations when they are genuinely trying to break with the past.

These unidentified craft are surveilling foreign military sites. The foreign countries have tried to shoot the craft down, and they've created internal organizations in order to lie to their own citizens.

In other words, other countries are pretty much acting just like the United States.

It sounds credible, but it's also incredibly convenient for the forces in our own country who've kept this information secret for decades.

▲ ▲ ▲

In his opening statement, independent journalist Michael Shellenberger said:

> One of Congress' most important responsibilities is oversight of the executive branch in general and the military and intelligence community in particular. Unfortunately, there is a growing body of evidence that the US government is not being transparent about what it knows about unidentified anomalous phenomena, and that elements within the military and IC are in violation of their constitutional duty to notify Congress of their operations.
>
> President-elect Donald Trump and former president Barack Obama have both said that the government has information about UAPs that it has not released. There are other current explanations for UAPs that they represent a new form of life, or nonhuman life. Current dominant alternative theories, including those put forward by AARO, are that UAPs are some kind of natural phenomena we don't yet understand, like ball

lighting or plasma. They could also be part of some new US or foreign government weapons program, such as drones, aircraft, balloons, CGI hoaxes, or birds.

Whatever UAPs are, Congress must be informed, as must the people of the United States. We have a right to know what UAPs are, no matter what they are. However, we now have existing and former US government officials who have told Congress that AARO and the Pentagon have broken the law by not revealing a significant body of information about UAPs, including military intelligence databases that have evidence of their existence as physical craft. One of those individuals is a current or former US government official acting as a UAP whistleblower.[321]

Shellenberger's testimony consists of much along this line, as well as trying to answer as best he could regarding relevant information about the still-anonymous whistleblower. One gets the sense in reading the transcript that some Congress members were frustrated by Shellenberger's appearance, as they were hoping to hear from the whistleblower.

On December 10, 2024, coauthor Kent Heckenlively was fortunate to be asked to drinks and appetizers with Michael by his good friends Jane and Joe Kearney of the Liberty Forum of Silicon Valley, California. Shellenberger was scheduled to give a speech at the Liberty Forum, and Jane and Joe usually take the speaker to dinner before the speech. As the group sat down, Jane observed that both Michael and Kent had easygoing, optimistic temperaments, and she expected the two to get along well.

Heckenlively's conversation with Shellenberger was pleasant. He tagged along as Shellenberger went to the reception before the speech, and Heckenlively sat next to him in the main auditorium as he waited to give his speech. The two talked about the recent national election and about Shellenberger's congressional testimony on UFOs. Heckenlively told him he'd written about the hearing and asked if he'd consider giving an interview, and Shellenberger agreed. The next day they recorded that interview.

Heckenlively began by asking him how he'd come to be interested in the subject. His response:

I started covering the UAP issue last year after David Grusch gave his testimony. I developed a number of sources who were able to help me report out a couple of different stories, including verifying or at least supporting the claims by Grusch that the US government had retrieved crashed craft of unknown origin. I then did a piece on claims of death threats and other threats being made against UAP witnesses and whistleblowers. I also did a story about the number of people that have come forward to share information, and their frustration and concern that the information wouldn't be used by the federal government, or that there wasn't the right system in place to protect whistleblowers.[322]

Heckenlively then asked him about the claim that the Immaculate Constellation whistleblower was expected to testify that day. Shellenberger replied that it must have been a misunderstanding on the part of a congressional staffer. He went on to discuss some of the more unusual parts of his recent testimony.

I think I've testified now over a dozen times [before Congress], and it's always been antagonistic between Republicans and Democrats. I've never been in a hearing where the Democrats and Republicans got along, much less seemed like friends. It was weird. There's a lot of strange aspects of this phenomenon, and one of the strangest is that it's bipartisan[...]. I think it's fair to say that the members of Congress feel lied to, because they have been lied to. So, they're annoyed with the Defense Department. They know they have a constitutional right and responsibility to provide oversight to every government agency, including the DoD and the different agencies in the intelligence community. They don't feel they're able to do their jobs. They don't understand why.

They also feel like they're being treated like children. My most viral comment was where I said, "We're tired of being treated like children." That's me, giving voice to a pretty widespread attitude among members of Congress. I think a lot of members of Congress think that UAPs are NHI.[323]

They talked more about the frustration of Congress members, then Shellenberger raised an idea Heckenlively hadn't considered.

I think there's a way in which the whole issue got framed which is actually bad for disclosure. It was framed as, "Do you believe it's aliens or not?" I don't think that's the right question. I think the right question is, "Why is the government hiding this from us? And what's it going to take for them to stop treating us like children?"[324]

The two then talked about the likely size of the black budget that has paid for these projects, as well as the problem of overclassification of government documents. Shellenberger believes there has been a genuine disclosure effort and that the whistleblowers are trying to play it straight.

Heckenlively next questioned him about whether the fear of government prosecution is justified. Would the deep state really prosecute people for talking about UFOs? Shellenberger has little doubt.

I think part of having this incredible military is also having a massive security apparatus that spies on its own people and monitors every single computer movement. For me, I don't have any doubt that they're genuinely scared. Because I talked to them, and I don't think they're faking that. I think there's real legal stuff there. I'm trying to redirect the impatience [about the whistleblowers] toward the target which needs to be pressured. Which is Trump, Kash [Patel, current director of the FBI], [John] Ratcliffe [current director of the CIA], and [Tulsi] Gabbard [current director of national intelligence]. These are the people that have to get the stuff out. I don't think any more pressure on the whistleblowers, or the journalists, or the members of Congress, is going to have any impact.[325]

If the deep state has been concealing this information for years, would it just roll over if a president, FBI director, CIA director, or director of national intelligence told it to comply? It's difficult to maintain the facade of a democracy if the public demands information about a subject and the government refuses to give a good explanation for why it can't provide it.

Perhaps the increasing public pressure will cause the dam to break, the information will be released, and we'll learn what the government has been concealing from us.

▲ ▲ ▲

Another unanswered question from the hearing is whether our government has been combining alien DNA and human DNA to produce hybrids with enhanced capabilities. From the transcript:

> **CONGRESSWOMAN LAUREN BOEBERT:** Okay, so there are rumors that have come up to the Hill of a secretive project within the Department of Defense, involving the manipulation of human genetics with what is described as nonhuman genetic material, potentially for the enhancement of human capabilities—hybrids. Are any of you familiar with that? Yes or no?
>
> **TIMOTHY GALLAUDET:** No, ma'am.
>
> **LUIS ELIZONDO:** I am not, ma'am.
>
> **MICHAEL SHELLENBERGER:** I'm not.
>
> **MICHAEL GOLD:** No, ma'am.[326]

Michael Gold is a former NASA official and currently the chief growth officer at a company named Redwire Space. He made it clear he was not speaking on behalf of NASA or Redwire. He testified last and did not seem to contribute much to the conversation, other than to state that many of NASA's sensors could easily be reconfigured to search for these objects. This was part of his prepared testimony:

> I am here today to speak out for science. Science requires data which should be collected without bias or prejudice. Yet, whenever the topic of UAP arises, those who wish to explore the phenomena are often confronted with resistance and ridicule.
>
> For example, members of the NASA UAP independent study team, particularly those in academia, were mocked and even threatened for simply having the temerity to engage in the

study of UAP. Our best tool for unlocking the mystery of UAP is science. But we cannot conduct a proper inquiry if the stigma is so overwhelming that just daring to be a part of a NASA research team elicits such a vitriolic response. Therefore, one of the most important actions that can be taken relative to exposing the truth of UAP is to combat the stigma. And that is where I believe NASA can be eminently helpful.[327]

It was certainly a noble sentiment, but seemed to be several years late.

While there was certainly some new information in the hearing, I couldn't help but feel a sense of frustration. Where was the smoking gun?

Heckenlively called Congressman Eric Burlison to learn what the committee members were thinking behind the scenes. His response:

What they said is that the impetus for the meeting was that there was a whistleblower that was coming forward. And that this person was connected to this document on a program called Immaculate Constellation. That's when I first heard of the term. I started getting questions from people that had heard about it. Asking if I'd seen the report. We requested that from Nancy Mace's office, if they had access to the report and could get us a copy. To my understanding, we weren't going to see the report until the hearing day, and that we would hear firsthand from the person who composed the report.

And unfortunately, that person did not come forward. Instead, it's a reporter. And I found that to be disappointing. This is the real cluster [mismanaged and chaotic situation] that occurred. On the morning of the meeting, we're in the back room—the green room, if you will—with all of the different witnesses, and I asked the question, "Where's the Immaculate Constellation document?"

There was confusion as to who had it and what the document actually was. The committee staff seemed confused, because they said, "Well, we have an eleven-page document, but it's just a Word file. We don't know if that's what you're talking about."[328]

We can't rule out the idea that such a monumental screwup indicates a conspiracy. But there's also an enormous amount of incompetence in any organization, including Congress. Conspiracy and incompetence can easily get you to the same place.

Burlison continued venting his frustration to Heckenlively about the hearing and the events surrounding it.

> The other thing that happened is, they were requesting witnesses. I've been receiving the names of individuals I think could get us closer to the information we're looking for. Some are likely to be hostile witnesses. They're individuals who don't want to be whistleblowers. Other people are outing their identity.
>
> And if we bring them forward and put them under oath, they'll probably tell us as little as possible until we get them into a SCIF. But at least I'm trying to get those names and have been submitting them to the committee. They ultimately put together a list of potential witnesses, and then asked us to rank which ones we wanted to have come forward in a hearing. Some of the people on that list of potential witnesses are, in my opinion, more interested in being on Netflix than having firsthand knowledge of anything.
>
> I want to talk with people who have firsthand knowledge. I'm done with all this hearsay. All this secondhand, "I was told" or, "I saw a report." No. All of these copies of copies of things are baloney. I don't know that we can trust it or understand exactly what we're looking at."[329]

Heckenlively then asked Burlison if he could provide his understanding of the government secrecy issue, particularly the role of Immaculate Constellation. His response:

> It sounds like there's an umbrella program that's called Immaculate Constellation. And it's an umbrella of several Special Access Programs that are designed to be compartmentalized so that no one really has all the information. It's designed so that they [intelligence agencies] immediately take whatever

information they have gathered and try to get it outside of the military's purview. So that even people within the military apparatus who have a "need to know" or should have access have nothing to look at. That's why so many people, when asked, have no information. They've never seen anything.[330]

Heckenlively then spun out for the congressman an explanation he'd heard from attorney Danny Sheehan. According to Sheehan, when the Roswell crash happened in 1947, President Truman established the MJ-12 group to determine intentions, reverse engineer the craft, and develop countermeasures against those craft.

In the document creating the MJ-12 group, Truman cited Article II of the US Constitution, giving him the power to conduct negotiations with foreign powers. Those who worked in MJ-12 would come to consider themselves the guardians of this knowledge, and eventually those in the inner circle would have what would come to be known as "Cosmic" clearance. The group having Cosmic clearance would consider themselves to be working under Truman's original grant of authority, which they believed gave them the right to treat future presidents as temporary occupants of the White House. Depending on the opinion of this group about the president, some presidents might be read into parts of the UFO issue while others would be shut out completely.

Burlison found this an interesting scenario, as it would explain why the efforts of some presidents as well as Congress had been brushed off.

The two continued discussing the possibilities of the MJ-12 group: its decades of secret work, shadowy individuals with Cosmic clearance, and all the various possibilities. Heckenlively finally told Burlison that he wondered if any of this stuff was real. Burlison responded:

> That's exactly how I feel about this. As I was reading this report, part of me was wondering, "If you're the US military, and you're spying on Russian bases and nuclear sites, don't you want to leak a report saying this is a global phenomenon?" That there are these vehicles that are being seen over military installations and nuclear sites and we don't know what they are? To me, it would be a very easy cover for some of the espionage we are doing.

Heckenlively then speculated about how the crash retrieval program could also be a perfect cover in case the military did shoot one of these things down, completing the entire loop of testing, recovery, and secrecy.

"That's right," Burlison said.[331]

The congressman and Heckenlively might not have the answer to whether aliens exist, but the two are convinced that powerful individuals in our government have been lying about the issue. Is it because aliens and their craft are real, and our leaders want to defend us from any possible harm? Or is it because there are no aliens or unidentified craft, and those in charge have created this fantasy as the perfect cover story for something else?

One would think that in light of all the recent disclosures, we would know more than we do. One thing seems clear: smoke and mirrors conceal the truth and make fools of all of us.

It's hard to stay sane and rational when we're all drowning in an ocean of lies.

CHAPTER ELEVEN

THE CAVE OF THE ALIEN MUMMIES

U p until this point, we've purposely avoided the question of whether aliens and UFOs have been visiting our planet throughout history.

We're aware of claims that certain events, like the spinning wheel in the Bible's Book of Ezekiel, the appearances of angels, and the appearances of beings from the skies as described by Native Americans or ancient Sumerians, were actually alien encounters.

Trying to validate and document questionable events from the past eighty years has already been such a challenge that we didn't want to go back thousands of years.

However, in September and November of 2023, there were two presentations in front of Mexico's General Congress by journalist Jaime Maussan that captured the attention of the world, as well as prompted quick denunciations that the alleged "three-fingered Peruvian mummies" (often referred to as tridactyls) were fakes. This is from a November 2023 Reuters article:

> Mexico's Congress heard from researchers on Tuesday who declared authentic a set of three-fingered Peruvian mummies recently presented as potential evidence of non-human life forms, while declining to certify that the remains were extraterrestrial.
>
> Lawmakers first heard from Mexican journalist and UFO enthusiast Jaime Maussan on Sept. 13 when he presented two specimens in a first-of-its-kind congressional event on UFOs, or FANIs in Spanish. Maussan said the bodies, believed to have been found near Peru's ancient Nazca lines, were not related to any life on Earth.

At Tuesday's session, Maussan was more focused on proving the bodies, which were not on display this time, were not fake, ushering in a string of doctors who all said the bodies were those of real, once living organisms.[332]

As with the hearings of UFO whistleblowers before the United States Congress, the presentations before the Mexican Congress convinced us that the story was credible enough to require coverage in this book.

We were also fortunate that coauthor Michael Mazzola, would soon find himself in the middle of the story, traveling to both Mexico and Peru to investigate. The Reuters article continues describing Maussan's testimony:

"None of the scientists say [the study results] prove that they are extraterrestrials, but I go further," he [Maussan] said, suggesting that they could be evidence of non-Earthly life forms.

Anthropologist Roger Zuniga of San Luis Gonzaga National University in Ica, Peru, said researchers had studied five similar specimens over four years.

"There was absolutely no human intervention in the physical and biological formation of these beings," he added, saying he didn't know the origin of the beings.

Zuniga presented a letter signed by 11 researchers from the university, declaring the same. The letter made clear, however, that they were not implying the bodies were "extraterrestrial."[333]

That story, combined with the fact that multiple researchers had investigated the claim, piqued our curiosity. Michael visited Mexico City from January 13 to January 16, 2024, to gather additional information about the claim. On January 17, 2024, Heckenlively recorded an interview with Mazzola to document what he'd found. He began by asking Mazzola to go through the backstory of the mummies and what he'd learned on his recent trip. Mazzola told Heckenlively:

These things were found in 2017 and dismissed as fraud. Then back in September of 2023, Jaime Maussan, an extremely famous journalist in Mexico, a correspondent on *60 Minutes Mexico*, who's been covering the UFO story for decades, walked these two alleged alien mummies into the Mexican Congress.

The story went totally viral but wasn't treated with much seriousness. People said it looked like [paper-mache] or whatever.

These things were allegedly found in a cave in Peru, a mass grave with dozens of bodies. The Peruvian minister of culture came out and said these are fakes: "They're just dolls made of chicken bones and modern glue. It's a hoax. It's a bad hoax. It's a hoax. You're ridiculous if you believe this." I just went to Mexico and interviewed one of the scientists who's been studying this.[334]

Mazzola then texted and emailed me nineteen photographs of what appeared to be the mummified remains of small gray aliens, from two to four feet tall. By the same token, the things in the pictures also resembled items constructed of paper-mache. (Some of these pictures are included in the center section of this book.) Michael continued with what he learned while in Mexico City:

These things were packed in this mass grave in diatomaceous earth. Whoever buried them knew exactly what they were doing, because they're perfectly preserved and they've got muscle and tissue, even their organs. Part of their brains are intact. That's usually the first thing that deteriorates. One of them is pregnant with three fertilized eggs in different stages of development. You can see these in the scans. [Author's note: I did see what appeared to be eggs in one of the scans.]

Dr. José Zalce [full name: José de Jesús Zalce Benítez, a forensic military doctor], who I interviewed, explained to me this is something that can't be forged. You can't fake something like this. Even the joints have wear and tear from use that you'd expect to see in a creature that lived and drew breath and walked the earth. And there are more bodies at the University of Ica in Peru.

They have twelve different labs which have independently replicated the DNA analysis. It showed that these things share 70 percent common ancestry with us. To give you some context, flies and humans have about 60 percent similarity in all genes, and about 75 percent in the genes that cause disease. With chimps and humans, we share about 98.8 percent of the same genes. I don't remember exactly what Zalce told me, but these beings had clear indications of being genetically engineered, as opposed to having evolved naturally.[335]

Heckenlively's thoughts were racing with this information. Was this smoking gun evidence of aliens? Mazzola continued his account:

And even if these things never walked on another planet, even if they're not extraterrestrial, they're dated at more than a thousand years old. Who had the ability to genetically engineer something over a thousand years ago? It had to be an extraterrestrial civilization.

One of these bodies that we filmed has a breastplate. You can see this sort of white wishbone-looking thing, but it's metal. It's made of cadmium and something else—I forget. This is over a thousand years old. We didn't even recognize cadmium until the twentieth century. This is not something humans could have done.[336]

Mazzola then spent a good deal of time talking about he and Steven Greer (with whom he'd made many films about UFOs) were going to get a scientific team together in the United States to examine the bodies and make sure the investigation was done correctly. Mazzola was enraged by the Peruvian government's attempt to dismiss the bodies as fakes.

What is definitely for sure is that the Peruvian government is full of shit. They are dismissing these as chicken bones and glue. If you Google "Nazca alien mummies," you'll see a bunch of stories from the last few days, and they put a dress on one of them. They put a little Mexican quinceañera [celebration of a girl's fifteenth birthday] dress on one of them to make it look stupid. And they admit it's a doll.

I've got the ability to put these things in front of members of the US Congress. But we can't move these things out of Peru without permission from the Peruvian Department of Antiquities.

But the Peruvian government has taken the position that these bodies are just chicken bones and string and glue.[337]

Mazzola raised another disturbing possibility, that these creatures could be genetically engineered beings who could have come close to passing as human.

Some of the scans are on another body that I didn't get to see personally. It's a much larger specimen called ["]Mary,["] and she looks very human, except she's got this elongated skull and very long fingers. It's three fingers, just like the little guys I was filming. Three long fingers and no thumb. But the fingers wrap—they coil around whatever they're trying to grasp, so they don't need a thumb.

And what's interesting is, her elongated skull was formed naturally. It wasn't formed the way that African tribes will do it by tying plates around a child's head to form the skull. But otherwise, she's very humanlike.

And something else interesting with the little guys. The way their necks are structured, they have the ability to elongate their necks in the way the creature did in [Steven] Spielberg's movie *ET*. The claim is that Spielberg was given access, because [for the making of] *Close Encounters of the Third Kind*, it's well-documented that they opened Project Blue Book for him. So, it makes sense that *ET* would also be based on a real event where one of these creatures was left behind.[338]

Could all of this information be out in the open, in Spielberg movies, giving us a hint of what's to come but letting us slot the information into the realm of the fictional, until we're able to accept it as reality?

What happens to our brains when we confront something that is truly alien, such as beings with long fingers and no thumbs?

⋏　⋏　⋏

On February 12, 2024, Heckenlively recorded another long interview with Mazzola. He'd recently returned from Peru, where he'd been gathering more information about the alien mummies. He was in the process of setting up a press conference in Los Angeles, which would also feature a live feed of scientists in Ica, Peru, performing CT scans on the bodies and discussing what they were seeing.

Mazzola first gave me background information about the endeavor and the work of Jaime Maussan, as well as the continuing battle with the Peruvian government.

> In Peru, I was filming these scientists, reacting for the first time to the autopsy, CT scans, and X-rays. And while we were doing that, Jaime Maussan was doing UFO stories, as he's been doing for decades. Remember, he's a household name in Mexico and Central America. [Michael then sent me a picture of the body they autopsied, which was a figure in a seated position, somewhere between eighteen and twenty-four inches in height, looking like a three- or four-year-old child but with an enormous head.]

> They did the autopsy and X-rays and everything. And they said it was a real being, but they don't know what it is. They have to do DNA on this one and the others in the new batch. But it's from the same tomb, the same cave, perfectly preserved in diatomaceous earth. So, it's mummified, and it's got organs and brain tissue.[339]

Heckenlively was intrigued to hear Mazzola's story of what had gone on with these manufactured dolls which being passed off by the Peruvian government as alien bodies. As Mazzola explained:

> They were made by a guy named Manuel. He told the authorities, "No, no, these are not the real ones. These are fake ones. These are works of art. I made these to capitalize on the tourism trade in my little town."

> They ignored this and claimed they did DNA analysis on the dolls and they were just chicken bones and glue. So, the Peruvian minister of culture is perpetrating a fraud on the public with these claims, and they've defamed the scientists.[340]

Heckenlively asked Mazzola to start from the beginning of his Peruvian adventure.

The first thing we did when we arrived in Lima, Peru, was to drive to Nazca, to look at the lines drawn over two thousand years ago on the desert floor. [The Nazca lines are large figures drawn on the desert floor that are visible only from the air.] Everyone's calling them the Nazca mummies, but they're not from Nazca. They're from Palpa, a little town next to Nazca that nobody's ever heard of.

So[,] the guys who found them have called them the Nazca alien mummies. We went and got into a little Cessna to fly over the lines. These lines are only properly visible from hundreds of feet up in the air. So, it begs the question not only of who made them, but for whom were they made? And we're posing the question, is there a connection between the Nazca lines and the mummies found nearby? Now, we can't possibly answer that question, but we wanted to pose it in our documentary.

But as soon as we got up in the air, I started puking my brains out, so I really didn't get a good look at the Nazca lines. I'll review the footage at some point.[341]

There were more thrills to come. What Mazzola wanted more than anything else was to not only document the location where these bodies had been found but get those who'd made the discovery on camera.

Then the next day we bribed the tomb raider [Mario] three thousand dollars in cash to take us to the cave. So, we had to drive out to the middle of nowhere. We arrived at a giant chicken coop and behind it was this mountain. We spent about two hours climbing the mountain to reach the cave. Some parts of it were a vertical climb and I thought we were going to fall. We filmed the entire thing.

And finally, we got close to the top, and there was a cave. I expected a cave I could just walk into. Nope. You had to get on your belly and squirm into the thing. I got two feet into this and started having a massive panic attack. I'm very claustrophobic. So[,] we sent the camera crew in. It was a very small cave.

The question we had is how all these bodies fit into this cave. But Mario explained that it was actually packed to the ceiling with diatomaceous earth. It was basically high enough and big enough for maybe four to six people to sit in.

It took Mario two years to excavate it with his hands. And he said that as a boy (he's in his forties now), he'd play in the area and see strange orbs in the sky over this cave.[342]

Some parts of this remarkable story made sense, but why would orbs be hanging around this burial site a thousand years after it had been sealed up? Heckenlively pressed for more details, such as the number of alien bodies that had been excavated. Mazzola replied:

I get different answers on how many bodies were there. And the reason I get different answers is because he's been illegally selling many of these on the black market. As far as I can tell, there were between 60 to 100 of them of varying shapes and sizes. But here's the interesting thing: while the cave is empty, it sounds like there's a hollow chamber on the other side of the wall. There might be additional chambers that could be excavated.[343]

Mazzola showed strong emotion again when discussing the Peruvian government's claims that these bodies were fake.

And the Peruvian government is just full of shit saying these are constructed of chicken bones and glue. You have other anthropologists who've looked at the original bodies. And they've put together reports saying, "Oh well, this must have been constructed." But even they say, "We have no idea how this was constructed, especially a thousand years ago. It's a complete mystery."

They say these were constructed, even though they have no evidence for it. They won't even consider the possibility they're real. They debunk themselves with arguments that make no sense.[344]

We ended our conversation soon after, as Mazzola was getting ready for that big press conference. Heckenlively was left struggling with all of this information and sought some verification. He found it in a December 1, 2023, *New York Post article*:

> Controversial journalist and UFO enthusiast Jose Jaime Maussan ushered in a team of scientists who performed a DNA analysis of the remains that found 30% of its genetic material is "not from any known species."

> Mexican researchers who appeared beside him as he testified before the nation's Congress also claimed the figures were "authentic," comprising of just a single skeleton, according to UK's Daily Star.

> While it remains unclear what the other 70% of the so-called aliens' DNA matched up with, Maussan argued that the DNA analysis confirms the mummies are "definitely not human."[345]

Much of this sounded like science fiction, but it was no less remarkable than the UFO whistleblower hearings held in our own Congress. The gap between skeptic and believer appeared to be shrinking dramatically. The article continued:

> "They're real," anthropologist Roger Zuniga of San Luis Gonzaga National University in Ica, Peru said of the specimens in his testimony last month.

> "There was absolutely no human intervention in the physical and biological formation of these beings," he added, saying he didn't know the origin of these beings.

> Zuniga also presented a letter signed by 11 researchers from the university testifying to the same conclusion—though they claimed they were not implying the bodies were definitely aliens.[346]

As Heckenlively searched for more information to support Mazzola's claim that they were trying to get various experts to validate the remains, he ran across another article from the *New York Post*. It read in part:

> Joshua McDowell, a former Colorado prosecutor and current defense attorney, examined one of the tiny, strange bodies—named "Maria"—with three independent forensic medical

examiners from the United States. [Author's note: in an interview with me, Joshua McDowell said that the *New York Post* got the name of the specimen wrong. He asserted that Maria was a large, near-human-size specimen.]

He and the experts were shocked to discover that the fingerprints on the ET-like corpses were in perfectly straight lines.

"These were not traditional human fingerprint patterns," attorney, Joshua McDowell, told the Daily Mail.

"We did not see any loops or whorls on the prints and fingers or on the toes," he said. "I'm a former prosecutor. I'm a criminal defense attorney. And these were not classic fingerprints," he added.[347]

Would the alleged forgers of these bodies have thought to put a weird, straight line of prints on the fingers and toes? Again, Mazzola's question of why somebody likely would have had to spend millions of dollars to create these items, then sell them for thousands of dollars, echoed in Heckenlively's mind.

McDowell and the US medical examiners traveled with Maussan to Peru last April to study the bodies. While the lack of human fingerprints is puzzling, he said it would be "extremely premature" to make any statements about the mummies' origin.

One possible explanation for the unusually straight fingerprints "could possibly have something to do with the way her skin was preserved," he said, noting that it's "very odd."

Maria's body is covered in diatomaceous earth—a type of white powder made from the sediment of fossilized algae found in bodies of water.[348]

What Heckenlively was finding in news accounts was checking out with what Mazzola had been telling him. The article provides additional information about McDowell and the American experts who accompanied him.

McDowell, a former deputy district attorney in Colorado's Fourth Judicial District, took a city coroner from Denver and a forensic anthropologist from Maryland's state's Medical Examiner's Office to the South American country to investigate.

He also brought his father, Dr. John McDowell, a forensic odontologist and retired professor at the University of Colorado, who assisted in identifying human remains after the 9/11 attacks on the World Trade Center, according to the Daily Mail.[349]

Heckenlively interviewed both McDowell and Zalce, and here's what he found.

McDowell grew up in Boulder, Colorado, where his father taught at the University of Colorado in both the medical and dental schools, prior to becoming president of the faculty. The younger McDowell went to law school at the University of Colorado, Boulder, worked for a year at a downtown Denver law firm, then spent three years working as a deputy district attorney in Colorado Springs before opening his own criminal defense firm.

The elder McDowell, prior to his trip to Peru with his son, received the prestigious honor from the American Academy of Forensic Sciences (AAFS) of being named R.B.H. Gradwohl Laureate. He'd also received multiple teaching awards from the University of Colorado, including "Professor of the Year, in addition to serving two terms as President of the Faculty at the Health Science Center, and three terms as President of the Faculty Senate for the University's School of Dental Medicine.[350]

Prior to becoming involved with the Nazca mummies, both McDowells had no experience with the UFO issue. But when Jamie Maussan presented his findings before the Mexican Congress, the younger McDowell was mesmerized. He recalled:

> At the end of it he [Maussen] says, "We welcome forensic experts from all over the world to come look at the bodies we have.["] And that's what my dad does. He's the former president of the American Academy of Forensic Sciences. He won the highest award in forensic sciences, the Gradwohl Medallion.

> So, I called him up and said, "You've got to see this." I sent him the link, and he watched it. He was intrigued by it as well. And I said, "Maussan invited experts from all over the world to come look at these. What do you think?"

And he replied, "I'll take a look. Why not?"

I emailed Jaime Maussan, and within an hour we were on the phone. And that's kind of how it went.[351]

The elder McDowell received the Gradwohl award at the end of February 2024, and it was a large family event, in addition to being attended by his academic and professional colleagues. The chief medical examiner for the City and County of Denver was an old friend, and the elder McDowell works with him as an assistant medical examiner. The McDowells invited him and a forensic anthropologist who works for the State of Maryland to join their investigation. Both men agreed, and the group of four made plans to fly down to Peru over the Easter weekend.

According to Joshua McDowell, the University of Ica is in possession of four alien mummies, named "Maria," "Wawita," "Alberto," and "Victoria." Other humanoid bodies named "Montserrat," "Sebastian," and "Santiago" have been recovered but are not currently located at the University of Ica. Maria, Montserrat, Sebastian, and Santiago are of varying ages and sizes. Alberto and Victoria are of the smaller variety. Maria and Montserrat are taller than the other two, estimated to be four foot eleven to five foot two. They're curled in a fetal position, so it's difficult to determine an exact height.

But they do seem consistent with normal human size, especially when considering that area of Peru and the fact they're estimated to be over a thousand years old.

Heckenlively told McDowell he'd read the article featuring him in the *Daily Mail*, in which he'd noted that the fingerprints of the bodies didn't have the typical loops, arches, and whorls of normal human fingerprints. He recalled:

I said, "The fingerprints are the least strange things on these bodies, but I'll happily talk to you about the fingerprints." Most of the fingers and toes were covered in diatomaceous earth, but some of them have been uncovered. And they're horizontal or diagonal. We did not see loops, swirls, or the arches you'd see in the traditional human fingerprint pattern in the fingers and toes. And I don't know why that would be.

Is it something to do with the way the bodies were preserved, and the skin has been dried and desiccated? I don't know. I'm not saying these bodies aren't modified humans. But I'm certainly not saying they *are* either.[352]

Heckenlively asked McDowell if he could tell me what struck him as most strange about them. He began by talking about the head shapes.

The head shapes are not consistent with human head shapes. But they are consistent with what we know of the Nazca and Paracas people from that time. We're talking about a time frame from 700 BC to 700 AD. We know they did head binding that resulted in cranial modification. Every single one of the bodies that we saw of the Maria type had cranial modifications.

Now, are those consistent with modifications that were done at the time of other Nazca and Paracas individuals? It's hard for me to say. We need better scans, and we need more time with them and comparison to others from that time period.[353]

The inevitable question that arises is whether the practice of head binding arose in an attempt to mimic something else, or as a novel concept in that area of South America without any outside influence.

However, what is unprecedented about the specimens is that they have three fingers (missing anything resembling a thumb and pinky finger), are extremely long compared with human fingers, and have an extra phalange, or joint. Similarly, the feet have three toes. "So, that is new," McDowell told me. "We don't have a fossil record of a humanoid body with three fingers and toes."[354]

Heckenlively mentioned how Mazzola had told him that the extra phalange allowed these fingers to act as a thumb, and asked if he had any thoughts about the claim. McDowell replied:

It's possible. I mean, we have to look at these things. A three-toed sloth has something similar. Have I heard that idea? Yes. But I'm agnostic on it. We have a living specimen of a three-toed sloth that we can look at to see how it functions and how it works. We don't have that here. Is this a real, biological, evolutionary, or genetic difference, or is it something that's been modified? We don't know.[355]

It seemed to Heckenlively McDowell was struggling to come up with a framework to describe something so far out of the range of typical human experience. At one point McDowell brought up the Fiji mermaid and the platypus. For those unfamiliar, the Fiji mermaid was a common sideshow attraction of the nineteenth century, a creature made by sewing the torso and head of a juvenile monkey to the back half of a fish. Most knew it was a hoax, but a lot of people still paid money to see it.

However, when the first stuffed example of a real platypus was sent from Australia to England for examination, many scientists thought it was a fake. The platypus is an improbable-looking creature, with the body of a small beaver, the bill of a duck, and poisonous claws. It lays eggs and excretes milk from its skin.

Throughout the interview, McDowell kept coming back to the question of whether we were looking at a Fiji mermaid or platypus situation, such as when we discussed the alleged giant skull that the tomb raider, Mario, also found at the site. McDowell had seen only a picture of the giant skull and was not impressed by it. However, during his trip to Peru, he questioned Mario about the find. As McDowell wrote in his blog:

> The only person who I have spoken to about the giant head who has seen it in person is "Mario," the man who discovered the cave. During my interview with Mario, I sensed he was uneasy speaking about the head and was seemingly surprised I knew about it. He confirmed it existed and said that he believed it was a real biological skull. He had looked inside of it, and seen the jaw, and other anatomical features. When I asked how big the head was, he mimed with his hands the general shape and size of the head. I would estimate what he showed with his hands to be about 18 inches wide by 30 inches tall. (Note: I'm assuming the estimates of an 8-meter-tall body is based on the size of the head in relation to an expected body size.)
>
> Seeing his unease in discussing the giant head, I moved on to other topics as I knew our time was limited and we had a lot to cover. I did speak to him later in private about the head and am curious to see if what he told me checks out as we learn more.[356]

If Mario's claim is accurate, in addition to aliens being two to three feet tall, we now have specimens that might be up to twenty-five feet tall. It's difficult to escape the suspicion that if the aliens created these creatures, they have few moral qualms about using for their own purposes life forms they've modified on the planets they've visited.

Heckenlively moved on to the ultimate question: whether or not McDowell believes these are alien bodies or perhaps a new species of human that has never been described in the scientific literature. McDowell answered:

> Let me think about how to answer that in the best way possible. Of all the specimens, there are things that are not consistent with the known morphology of other specimens. It comes down to, either these are entirely new creatures or they're hoaxes. As I said, are we looking at a Fiji mermaid or platypus situation?
>
> The larger bodies, Maria, Montserrat, Sebastian, and Santiago, all have three fingers and three toes. And they have extra phalanges, the finger bones, in each of the fingers and toes. So, they're missing a thumb and a pinkie. And then the fingers are extra long because there's an extra phalange. Well, that doesn't make sense.
>
> Do they look very good under CT and fluoroscopy? They do. They look great. But keep in mind, these are covered in diatomaceous earth [which gives them a paper-mache-like appearance]. So, it's very difficult to get a good visual inspection on the external portions of the specimens, because they're covered in this thick, clay-like substance.
>
> So, that's the big question, right? Has there been tampering or not?
>
> And if there has been tampering, has it been covered by the diatomaceous earth? Because it almost looks like a cast. But it's been cleared off in places, or fallen off, and you can see fingerprints as well as skin. And it *is* skin. There's no question about it.[357]

While presenting all the evidence that he thought supported the idea that these specimens were living creatures, McDowell also went through a litany of the things he thought had not been done to fully support that conclusion.

He did not trust the DNA data showing that the samples were only 70 percent human, because they had not been taken from the teeth or long bones, as would be ideal for ancient DNA, and they had not been taken in a properly sterile environment. He believed that some of the DNA test results showed contamination.

He also believed that any investigation needed to be done with the full cooperation of the Peruvian government. Others had allegedly taken samples out of the country for testing, but this set up issues of chain of custody as well as of stealing the cultural patrimony of a nation. As McDowell explained:

> The big problem is international law treaties. There's a memorandum of understanding between the United States and Peru that says you're not allowed to bring anything that's pre-Colombian or pre-Hispanic out of the country. Anything that is considered an archeological artifact or ethnological material is designated as cultural property and can't be removed from the country.
>
> Essentially, you would need the Ministry of Culture to authorize any sort of transport and then[...] US Customs to verify. I don't know if the Peruvian authorities are in favor of allowing these things to be transferred, but they have not given permission. Their position to date has been that the bodies, or the specimens, are hoaxes and forgeries, and don't need to be tested any further.[358]

Heckenlively pointed out how that is a contradictory position. If they're hoaxes or forgeries, they certainly aren't more than 250 years old, so could not be considered the cultural patrimony of Peru. He replied:

> One would think. But as a lawyer, you know how that works, right? They get to have their cake and eat it, too. And they're not going to allow anything to leave the country. In fact, they've tried to seize the bodies on many different occasions. [The specimens are] at the university right now, and the university has autonomy, as I understand it. That's how it's been explained

to me. I don't have an independent legal basis to verify that this is true.

That's how it was explained to me while we were at the university. The university is autonomous, and [the government] [doesn't] have the ability to seize the bodies from the university. And as for the other bodies, I don't know where they are. But I think they would try to seize them, even though they say they're fake.[359]

As to what McDowell and the three experts who accompanied him to Peru think should be done, this is what he said:

If I was to give a summary of the doctors' opinions, as well as mine, we all agree that there needs to be further testing to determine what in fact they [the specimens] are. Everyone agreed that number one, we should have better imaging modalities. It was explained to me that their available CT scans in Peru could be analogized to a standard-definition TV, but now we have 4K. The United States has machines that are far better than what they have in Peru.

But these machines cost hundreds of thousands of dollars, and it's not practical to transport one of those to Peru. It'd be a lot more practical to simply transport a body.

Number two, we would want genetic studies to be done, specifically DNA taken from various parts of the body. With each specimen, the idea would be to sample different parts from a single body to make sure it is in fact one creature, one biological specimen—that they're not mishmashes of different creatures and body pieces.

And also, to see what they are. Are they human? Nonhuman? Do they share DNA? The things that have been done in the past—and there's been a lot of really good work, a lot of really great doctors, but it hasn't necessarily always been up to the standards that our doctors would have liked to have seen.[360]

It seemed that McDowell and the American team had come up with a plan to genuinely answer the question about whether the mummies were human or not. Heckenlively said it seemed that the Peruvian government was blocking the investigation. McDowell agreed, saying, "Yeah, it's a real Catch-22. It's like they're saying, 'They're fake, but you can't examine them.'"[361]

When Heckenlively mentioned to McDowell that he seemed to be open-minded about the specimens, and would have little hesitation saying they were fakes if that's what the tests revealed, he replied:

> That's right. I want to see the DNA. I want to see the carbon-14 tests. I want to see those high-definition scans. And if they are fakes, I'll be the first in line to say, 'This is BS.' But if you're going to say it's BS, you don't just get to say it's BS. You've got the have the documentation to prove it—the science and the evidence. And it goes both ways. That's why I've been very careful in talking to people to say, 'I don't know the answer.' The reason I don't know is because the tests haven't been done the right way.[362]

Heckenlively was still keeping an open mind about the Nazca mummies, but wanted to get a more credible scientific source. That meant talking to Dr. Zalce.

▲ ▲ ▲

José de Jesús Zalce Benítez, the military forensic doctor who examined the specimens, began his interview with Heckenlively on July 31, 2024, by apologizing for his poor English. During the conversation, there were only a few times when Zalce paused to find the right words, and just a single occasion when he couldn't find the word he wanted. Heckenlively did correct some of his grammar in the interview excerpts below but kept the substance of his remarks unchanged.

Mazzola later told Heckenlively about the trouble Zalce had encountered because of his investigations. At one point, Zalce was told by his superiors in the Navy to write a letter saying the mummies were fake and that Jaime Maussan was a charlatan.[363] When he refused to comply, he was thrown in jail for five days. Under a change of leadership, Zalce was allowed to resume his work for the Navy, but his work on the mummies had to be kept completely separate from his Navy work. Additionally, Zalce told Mazzola that his endoscopy and CT scans of the mummy known as Montserrat was able to determine with "99.9 percent certainty" that the fetus in its belly was also tridactyl, with three fingers and toes.[364]

Zalce began his interview with Heckenlively explaining he'd studied medicine at the Navy Medical School in Mexico, then received a master's degree in forensic science from the Mexican Army. He has worked for the Mexican Navy for twenty years and used to be the head director of its Forensic Science Services. He also has taught forensics at the medical school level and for postgraduates specializing in forensics for the Navy. According to one article, Zalce was currently director of the Health Sciences Research Institute of the Secretary of the Navy.[365]

Heckenlively asked how he'd come to work on the bodies, and he said that journalist Jaime Maussan had asked him in 2017 to go down to Peru to view the specimens. Initially, Zalce was skeptical, but explained how his thinking changed:

> But when I got closer and could touch, feel, and made X-rays studies and radiological studies, I was finding many things, and that changed my mind and made me think it could be real.
>
> When we took X-rays, the first thing we found was a perfect junction in the joints between the bones. That was something that was good for the authenticity of the bodies. And when we made the CT scans, we found many structures that would be difficult to falsify. That's what changed my mind to thinking it was real. But we need to do more analysis of the bodies.[366]

Heckenlively hadn't considered something as simple as the joints of these creatures fitting together as something which might be evidence of authenticity. Zalce continued describing factors that seemed to indicate the body was authentic and not a creation like a doll, as alleged by Peruvian authorities.

> The second thing was, in the CT scans we could see small structures, ligaments, tendons, and vessels.
>
> And the third thing we saw in the small body called Josefina was three eggs. When we did the CT scan, we located a fourth egg in the oviduct. That is impossible to replicate by a hand-

made forgery. These are some of the things which convinced me it was real in 2017. Through the years, we've found many other things which confirm to me it's real.[367]

Michael had told Heckenlively about the eggs revealed by the CT scan and had even sent a picture of the scan. But hearing this account from a forensic specialist gave it a new level of reality to Heckenlively. It seemed to Heckenlively the best evidence was the DNA findings, so he asked about that—specifically, the claim that the bodies share 70 percent of the human genome. Zalce replied:

> That's correct. Seventy percent is pretty similar to the structure of human DNA that we know in science. We don't know how close this DNA is to humans. We can only specify that this 70 percent is pretty much similar to the human DNA we know in science.
>
> The relevant thing is the 30 percent that doesn't match with human DNA. Because if you know about humans and the primates, we have more than 99 percent in common. With bacteria and viruses, we often share more than 85 percent of our DNA. So that 30 percent difference is huge in evolution. That's the relevant thing, this difference.
>
> Dr. [Ricardo] Rangel [full name: Ricardo Rangel Martínez, a molecular geneticist] did the DNA analysis. In addition to the human DNA, he found that there are also two types of apes, the bonobo and chimpanzee, as well as human DNA from the south of Asia.[368]

That was about the extent of what Zalce could tell about the DNA, and he suggested Heckenlively contact Rangel to discuss it in greater detail. Heckenlively then asked about the quality of the evidence was whether the mummies were over a thousand years old.

Zalce said that the carbon-14 dating showed an approximate age of around 1,700 years, meaning these creatures had been alive in South America around the fourth century AD.

Heckenlively next asked what could be said about the likely habits and capabilities of these organisms as compared to human beings. Zalce answered:

> By analyzing the structure of the bones in the hands and feet, we can say they were very strong and very fast. Their biomechanics are different than ours. But we believe they could move very fast.
>
> Their fingers are very long, and they often have two to three more phalanges than we do. Human fingers have three phalanges. [The extra phalanges] permit [the creatures] to hold whatever they want without needing a thumb like we do.
>
> But we need to be careful, because different specimens have different numbers of phalanges on the fingers. Some have only one additional phalange, while others have three additional phalanges.
>
> We need to do more study of the hands and feet. For example, the last phalange on the toes is always in a down position, like the claw of an eagle.
>
> The structure of the feet is like an arc. It makes me think that that mechanics for how they walk are very different than for humans. They have the capability to make longer steps and faster movements, because they have parts that allow them to make more adjustments while taking those steps. It makes me think that their mobility and speed are probably superior to that of a human.[369]

The next area Heckenlively questioned him about were the eyes. If you've seen pictures of the typical gray alien, you know that the eyes tend to be larger in proportion to their faces than those of a typical human being. Zalce said:

> First of all, they have a bigger size of this space for their eyes. If they have a bigger space, and the structure is at the front of the face, it's likely the eye had a wider field of vision.

Humans can see anywhere from 120 to 140 degrees of a field of vision with our eyes. In 2017, I estimated that these creatures likely had anywhere from a 180- to 200- degree field of vision.[370]

Heckenlively asked whether the placement of the eyes in the front of the face meant that these creatures were more likely to be predatory in nature. (It's a common trait of carnivores, unlike herbivores, which tend to have eyes on both side of their face, like a horse.) It was a difficult question for him to answer, because while the placement of the eyes suggests the creatures had evolved as predators, the teeth told a different story.

Everything you say about the eyes is correct, and if I considered only the eyes, I would agree. But when we look at the teeth, we see [that the creatures] can eat meat and plants. Probably more plants, and that they are likely omnivorous. In one specimen, Montserrat, there is material in the colon. I have not sampled that, but if we could, we would know more. The smaller specimens have no teeth, so we think they ate soft food.[371]

The eyes did seem to suggest that at some point in the distant past, the creatures had been hunters. But when one looked at the teeth, as well as other factors, the picture became more complex. Heckenlively kept coming back to an idea he'd encountered earlier, specifically in the work of Philip Corso detailed in his book *The Day After Roswell*. In examining the Roswell bodies, Corso concluded that they were "engineered" creatures.

Corso believed we were not looking at genuine aliens from another planetary system, but rather, creatures designed to perform the bulk of interactions on our planet. Perhaps in this age of genetic engineering and tampering with deadly superviruses, it's now a more believable scenario than it might have been seventy-five years ago.

Heckenlively mentioned to Zalce that the noses of these creatures appeared to be small, and asked whether that meant the sense of smell was likely not a robust sensory system.

Well, I cannot respond to that, because the capability of smell depends on the number of specific cells in the nose and the brain. The size of the nose only allows me to say that they were not able to breathe as we do. They need to breathe more quickly—

superficial breathing. The mouth and the nose would need to work together, with their mouths open to breathe, unlike us, where we can usually breathe just through our nose.[372]

And finally, Heckenlively asked if there were any other features of the bodies he hadn't asked about that Zalce considered important. Zalce said two: the fingerprints and the skin.

The fingerprints, because they are completely linear and parallel, which is not like any kind of primate we know, who have waves in their fingerprints.

In the bigger bodies, the skin is without hair, and we don't see any kind of hair anywhere on the body. And the skin is very soft, like you'd expect for a human. But in the small bodies, they have skin which is similar to a frog.[373]

The strangeness of these creatures seemed to have no end.

Heckenlively sought to get greater clarity on the DNA results from molecular geneticist Ricardo Rangel Martínez. Rangel's presentation before the Mexican Congress on November 7, 2023 was summarized in the following manner:

The massive sequencing of Maria shows that it would have been genetically designed [approximately] two thousand years ago.... Maria would be a hybrid of *Homo sapiens* (30.22 percent of genes), of chimpanzee, of bonobo (4.7 percent of genes of these large African primates) and an unknown relative. Maria would thus have four different types of genes. It is impossible that this is some sort of natural mutation. Someone must be at the origin of this genetic manipulation, around 1,750 years ago. The biologist speaks of a complex process of artificial hybridization.

Maria might be a male. It is difficult to determine this with certainty. This difficulty can perhaps be explained by hybridization. Maria's bones appear robust, like those of great apes.[374]

Heckenlively was trying to make sense of the various claims he'd heard. Were these specimens 70 percent human, or did 70 percent of their DNA pertain to the DNA of known earthly life (of which the majority is human, mixed in with great apes as well as some microbial contamination)?

The 30 percent refers to DNA that does not belong to any known life on Earth. That's the way Heckenlively resolved it in his mind.

Heckenlively kept thinking of the formulation of attorney Joshua McDowell: Were we looking at the Fiji mermaid or the platypus?

Heckenlively wanted to get his hands on the DNA report.

⋏ ⋏ ⋏

Michael Mazzola got in touch with Ricardo Rangel Martínez, who on August 6, 2024, sent Heckenlively an email with the DNA report attached. The email reads:

> Dear Kent Heckenlively,
>
> It is an honor to be able to share with you such important information. As you know, the three-toed mummies of Nazca have drawn the attention of the general public, and it is not a surprise, since the information that their anatomical and genetic study is yielding is placing us in front of a very profound conceptual paradigm.
>
> I share with you the studies and preliminary results that have been yielded by the nucleotide sequences obtained from these three-toed mummies. We hope, and it is the wish of the scientists involved in the study of these specimens, that what is expressed here can be disseminated, in addition to being able to invite scientists from around the world who have knowledge of bioinformatics to confirm the results that have been obtained up to this point.
>
> Thanking you in advance for your attention to this matter.
>
> Sincerely,
> Biologist Ricardo Rangel Martinez.

The report Rangel sent me consists of two sets of data: a DNA analysis of an individual referred to as Victoria and test results by Canadian researchers. The opening of the report contains a statement Rangel made before the Mexican Congress on September 12, 2023:

Continuing with the analysis of the biological entities referred to a few moments ago by Dr. Zalce, I have come here freely and voluntarily to give my interpretation of the studies carried out on the tissue samples obtained from the neck and hip of the individual called "Victoria," which was subjected to cleaning and extraction for DNA (deoxyribonucleic acid), to later carry out a DNA sequencing process, this in order to be able to identify the organism to which the specimen in question could be related.

Since these are samples of ancient biological tissues, which could be verified through carbon 14 tests, it was expected that the molecules that protect the genetic information of this individual could be altered, degraded or partially destroyed, this due to the effect of stress caused by multiple environmental variables.

Furthermore, it must also be considered that since the specimen was found in a cave, where it could have been in direct contact with microorganisms, insects, pollen, or any other living organism, the possibility that the DNA sample isolated from the aforementioned tissues could have been contaminated is very high.[375]

As Heckenlively read through the report, he could understand why McDowell believed that additional work needs to be done. The results were intriguing but in many ways incomplete. As specified in the report, the results of the first set of tests were:

    a.   From the sample of the neck bone tissue identified as WGS Ancient002, 72.07% of the reading sequences were identified and 27.93% of the reading sequences obtained did not match the genomes of living beings known to date.

    b.   Of the 72.07% of the readings identified, 70.45% belong to contaminating DNA sequences from Homo Sapiens and the remaining percentage belongs to viruses and bacteria that also contaminated the sample.

c. From the sample of muscle tissue from the hip of the specimen identified as WGS Ancient0004, 36.28% of the reading sequences were identified and 63.72% of the reading sequences did not match the genomes of living beings known to date.

d. Of the 36.28 of the identified genomes, all turned out to be contaminating DNA from contemporary viruses, bacteria, and plants, and the genome of no mammal, including humans, could be identified.[376]

Heckenlively was beginning to understand how the claim that these mummies were 70 percent of earthly origin and 30 percent something else had started. It is the clearest way to state the evidence, but it might also be mistaken. One could claim that the first sample (Ancient0002) is consistent with this characterization, but the second sample (Ancient0004) is clearly not.

From these two samples, Rangel made the following five conclusions in his report:

1. The two ancient biological samples contain contaminating DNA that could have been acquired during the manipulation of the specimen or over time, by organisms that surrounded it and by the people who manipulated the specimen.

2. The sample identified WGS Ancient0004 showed the highest number of unidentified sequence readings, about 316 million readings.

3. This high percentage of unidentified readings with the living organisms that we currently know should be the focus of future research to try to define the identity of the organism in question.

4. There is a probability greater than 90% that this organism is not related to humans.

5. There is a probability greater than 50% that this organism is not related to living beings known to date on our planet.[377]

Rangel lists additional steps that he believes should be taken, including further genetic testing as well as the creation of a multidisciplinary scientific committee to continue the study of the specimen.

The next section of the report is from testimony that Rangel gave before the Mexican Congress at its second hearing on the UFO issue, on November 7, 2023. This testimony detailed the results of a genetic study of the specimen known as Maria, which was given the designation Ancient0003. This set of results appears to be more robust than the previous test results, with contamination being at a minimum, as Rangel reported to Congress.

Of the 97.38 percent of reading sequences identified, a small percentage corresponds to DNA of microorganisms that contaminated the sample or that are part of the tegumentary microbiota—that is, 2.13 percent of the reading sequences isolated from the sample obtained.

Therefore, the remaining 95.25 percent correspond to DNA from the tissue obtained from the organism in question. As can be seen, there is a downward distribution in the percentages identified, as we move away from the ancestral lineages in the phylogenetic tree, until we observe the lowest percentage in the organism that is identified to genus and species....

When paying attention to the final results, it stands out and draws attention that 30.22 percent of the reading sequences are identified with the human genome, that is, *Homo sapiens*, but it also includes an anomalous taxonomic branch of the *Pan* [chimpanzee] genus, with 3.05 percent of the reading sequences obtained from the organism under study....

Surprisingly, you can see that the new information displayed corresponds to the reading sequences identified in two species of apes, which are identified by genus and species, being *Pan Troglodytes* and *Pan Paniscus*—that is, the chimpanzee and the bonobo [the latter is a smaller, less aggressive species of chimpanzee], respectively. Both species are sisters.

These apes are only found on the African continent, so the origin of this species on the American continent more than 1,700 years ago is a real mystery, which, according to the results of the sequencing, would have hybridized with a specimen of the genus *Homo sapiens*.[378]

These creatures are part human, part chimpanzee, and part something else, like in some bad science fiction movie. These newly discovered creatures might have the intelligence of humans combined with the strength of apes.

The DNA sample shows that the genes share a 75.10 percent similarity with the class *Hominidae*—the four species of so-called great apes—which includes orangutans, gorillas, chimpanzees, and humans, and an 86.68 percent similarity with the larger clade *Hominoidea*, the larger class of so-called upright primates, comprised of twenty-eight species.

Further on in the report is the following claim: "To our surprise, we managed to identify several experiments in which several hybrid neuronal cells between Homo Sapiens and Pan troglodytes were obtained."[379]

The implication was clear: the aliens not only were mixing their DNA with that of humans, but also were adding the DNA of near-human primates.

Rangel suggests that these specimens should be given the scientific name *Homopan tridactyla*, or in essence, the human-chimpanzee with three fingers/toes.

Further testing of the human portion of the DNA by Canadian researchers suggests that the DNA of a male human from the Chinese population was mixed with the DNA of a female human from the Myanmar (northern Thailand/Burma) region.[380] In other words, these Peruvian mummies from around the fourth century AD do not show human DNA from ancient Peruvians, but from ancient Asian populations. As reported in the discussion section of the Canadian researchers' report:

> Along with varying alignments (ranging from 10% to 95%) with the human genome, we discovered a wide range of ancient microbial life and an unexpected genetic lineage to Myanmar, specifically through the mitochondrial DNA haplogroup "M20a" in the sample Ancient0003. The lack of American subclades and the Asian lineage proposes new questions about the origins of the human DNA in the sample. Was it the most likely explanation—a simple mishandling of the samples along the

chain of custody—or does it enrich the story of human migration? How did this DNA thread from Myanmar weave its way into a Peruvian mummy? This deviation from the anticipated American lineage invites a wealth of questions regarding human migration and potential intercontinental connections in antiquity.[381]

One gets the sense that the Canadian researchers were twisting themselves into pretzels to try to explain the findings.

If one were simply examining a 1,700-year-old human-looking body found in South America, and the DNA revealed an Asian origin, it might make sense to question what we know of ancient migration between the two continents. However, that doesn't consider the strangeness of this specimen, the DNA contribution from two species of chimpanzees from Africa, the three fingers on the hands, the three toes on the feet, the peculiarities of the skull, and the other irregularities.

The bulk of the evidence provided to date on these specimens suggests that these are genetically engineered beings.

Somebody was performing experiments, and creating creatures, either that are beyond our current capability to create or that no sane scientist would ever attempt to create, because most people would perceive the resulting beings as monsters.

We have evidence of these creations.

What we lack is evidence of their creators.

⅄ ⅄ ⅄

On November 9, 2024, the Peruvian government held another "public audience" regarding the Nazca mummies, and invited attorney Joshua McDowell to make a presentation. McDowell explained to Heckenlively that in Peru, a public audience is akin to a US congressional hearing, and there were three Peruvian congressmen who were spearheading the effort.

When Heckenlively interviewed him on November 25, 2024, he was upbeat and optimistic about the idea that the Peruvian authorities were taking a closer look at the bodies. McDowell was particularly pleased that the authorities were now referring to the specimens as "the mummies encountered in the jurisdiction of Nazca." The purpose of the hearing was to get permission for the bodies to be tested outside Peru.

McDowell originally was given ten minutes to make his presentation. Then it was cut down to five, then three.

He related that the bulk of his brief remarks were in response to Flavio Estrada, a government anthropologist who seemed to have been chosen as the resident skeptic, and to pharmaceutical chemist Ernesto Ávalos. McDowell said of Estrada:

> He presented these dolls that were confiscated at the airport. And tried to show the Peruvian Congress that these dolls were the mummies. And that the serious scientists who had been studying the bodies were a joke. And he tried to make a joke out of the whole hearing by showing these dolls. And he tried to say, "These are the bodies. They're constructions. They're fake."

> They are not the bodies that we've been studying. They are not the bodies we've seen under X-ray and CT scan. They're completely different. They have nothing to do with each other. They don't even really look alike.

> And I said in my testimony that he was doing a disservice to the people in Peru and to his profession, because he wasn't being honest with the facts. And it's disappointing to see, because we have something that's unique and important and needs to be studied.[382]

McDowell went on to state that he was making no claims as to what the bodies were, simply that they were specimens that demanded further testing. He told me:

> We should get some DNA from a tooth, from a long bone, something of that nature. I wouldn't want to take skin tissue, because we're talking about ancient DNA. We'd want to take something from inside the femur, or potentially from inside a tooth, to be able to get that good, clean DNA that we could run for ancient DNA testing. Are they humans that have a genetic defect? Are they an unknown species?[383]

He talked at length about the new minister of culture in Peru, who had taken a much more respectful approach to the Nazca mummies, and in particular, the University of Ica, which possessed several of the bodies. The changed approach resulted in an agreement between the University of Ica and the Ministry of Culture for two of the bodies, Maria and an infant-size body, Wawita, to be taken to a local hospital for X-rays and CT scans. McDowell said, "At the Congressional hearing, Jamie Maussan did present a short video of two of the doctors who examined Maria. And they indicated that they did not see in the imaging any modification of the bodies, such as removal of toes or fingers, or adding extra phalanges."[384]

Heckenlively found the imaging to be interesting but was still trying to nail down what the genetic testing had revealed. McDowell had many concerns about the reliability of the testing that had been done to date. He said:

> The genetic testing is interesting, because everything's been done without the approval of the Ministry of Culture. And you have to look at how the samples have been taken. So, we have to talk about collection, contamination, and also chain of custody.

> Then we have to look at the testing methods that were used. Did they have the ability to do ancient DNA? It's one thing to take a swab of your cheek and send it into 23andMe. That's cheap and easy. But to do the right work on these bodies that are 1,500 to 2,000 years old, or possibly more, it's a whole different process. It requires much more sophisticated techniques and technicians as well.

> People have taken samples all over the place, even to Russia. I know where at least two samples are in the United States right now. But I have major concerns about all those things because of issues with chain of custody, collection, contamination, techniques, and testing. Because there's so many different specimens, different exemplars of bodies—like the ones called insectoids, which have these small sixty-centimeter-type bodies. Then we have giant skulls, which I've seen [in pictures].

> And I don't want to talk about those too much, because I haven't seen them in person. We haven't physically examined them, but

they exist. I have probably three dozen pictures on my phone of photographs and imaging modalities of these giant skulls.[385]

Just when Heckenlively thought he was getting comfortable with the idea of three-foot-tall aliens and spacecraft, Josh had to throw in insectoids and giants. McDowell seemed to be similarly troubled, saying that he had a hard time believing that these insectoids were ever living creatures, although he did admit that they sounded like the fairies, elves, and other tiny people of folk tales.

McDowell seemed more comfortable talking about those bodies of near-human size, their curious morphology, as well as the metal implants found in several of them. He sent Heckenlively a picture, which he opened on my phone, and then explained the image.

> It's shaped like a barbell, and it's metal. Some tests have stated that it's made of osmium, which is a platinum byproduct. I cannot independently confirm that. I've not seen the metallurgic analysis done on it, but I've been told they're made of rare metals. That's something I would like to do at a later date.

> Some of them have implants in the backs of their necks, or in their chests, that are of different metals. A lot of them have implants in their forehead as well.[386]

It was clear that McDowell felt passionately about the Nazca mummies but also was troubled by the chain of events that had led to their discovery.

> There've been a lot of artifacts found in the caves along with the bodies as well, carvings of various things and little metal plates with inscriptions. And it's hard to know what came from where, because the grave robbers went into the location, this cave, and removed everything. And we don't have the site for archeologists and experts to go and examine, because the grave robber has changed his story multiple times. And maybe that's in his best interest, keeping him from being prosecuted but also to throw off competitors, other grave robbers.[387]

Heckenlively asked how many anomalous bodies he thought had been discovered. McDowell replied that he'd personally seen around fifteen bodies and photos of fifteen to twenty more bodies, but had been told there were about a hundred in total.[388] He also told Heckenlively that there were many morphologies which hadn't yet been publicly disclosed.

As they were winding up the interview, McDowell said:

> It's an ongoing story, and it's very interesting. There are so many weird aspects to this that when you think, "Okay, I've got a grip on this," something new shows up and really changes your understanding of what's going on.

> I personally think that if we can do some initial testing on the most human-like bodies—the Maria-, Montserrat-, and Fernando-type bodies—and get the genetic code, understand their genome better, we can learn a lot about what these are. Because we can also carbon-date them to understand exactly how old these are. Do they match anything we know? Are they relatives of modern-day Peruvians? Are they from somewhere else? Are they even 100 percent human? That's what we have to figure out.[389]

Time and again, we're told that the truth can be found just over the next hill. But when we reach the summit and see the valley below, we're confronted by yet another series of hills, with the Promised Land beckoning just a few more miles ahead.

If McDowell had personally seen somewhere in the neighborhood of thirty bodies, and supposedly more than a hundred had been uncovered, it seemed only a matter of time before the proper tests could be conducted and we'd have a definitive answer.

We were on the hunt for something, and whether it was aliens, human lies, or some combination of the two, nobody could say.

⅄   ⅄   ⅄

The final interview for this chapter was Serena DC (De Comarmond), an Australian documentary filmmaker who works with Michael Mazzola. The two of them have been investigating the Nazca mummies for the better part of a year and have shot a great deal of footage for a documentary tentatively titled *This Is Not a Hoax*.

Serena said that she and Michael had been following the story for some time and then received an invitation from Jamie Maussan to go examine the bodies in Mexico City. She told Heckenlively:

> We went to Mexico and weren't wildly optimistic that we would think they were extraterrestrials. We thought they were probably really, really well-made fakes. But then when we saw them, everything changed. To be close to them—and I mean, we got to physically touch them; we were that close—it became evident very quickly that if they were fakes, they were absolutely brilliant fakes.

> Then we were invited to go and see forensic tests being done on these beings. We got to be present when they were X-rayed, and when they had CT scans and MRIs. I can't explain to you how incredible it was to be in a room with scientists watching the imagery loading in real time, then seeing the archeologists and scientists burst into tears. Grown men in their fifties bursting into tears because what they were seeing was literally a new species. It was incredible.

> The interior of these beings is incredibly complex, just as complex as a human. Their skeletal structure has hundreds of bones. And what we found really interesting was—as you know, in forensics, if you find a dead body and it's badly decomposed, it's hard to know how old that body was by just glancing at it, because a lot of the flesh and things like that have fallen away, right?

> But they're able to look at the wear and tear on bones, specifically on elbows, knee joints, and ankles. And that's one way they're able to ascertain how old a person was when they died, because they could see the wear and tear and can compare that to what would be typical for someone of that age. This was something the scientists did with these beings. Not only were they proven to have walked; they had wear and tear consistent with living for at least three years. I found that really interesting as well. I can't imagine that someone making a fake would

consider that and then have the ability to mimic three years' worth of movement on a synthetic bone.

The plot kept thickening. More scans were done, biopsies were taken, and we were able to do DNA testing. It revealed that there were multiple earthbound animals that made up the bulk of the DNA, but about 30 percent of it was unknown. So, not only were we looking at a highly sophisticated modelmaker, but also a model who knows how to age bones, and a modelmaker who knows how to somehow implant mysterious otherworldly DNA into fake flesh. And the more we dug into it, it just became harder and harder for us to consider they were fakes.[390]

De Comarmond was charming to talk with. I'd talked with several cautious people in the process of writing the book, so it was refreshing to talk with somebody who made no secret of her passionate belief in UFOs.

De Comarmond went on to tell her version of how all those bodies came to be found in the cave.

Normally in tombs, their bodies are wrapped, prepared; the bodies are lying with their arms crossed over their chest—that kind of stuff. If someone's taken the time to bury you or put you in a tomb, they're not just going to throw your body in there and walk away, right? They're going to place you accordingly.

But that's not how these beings were. They looked as though they were hiding in this cave. Like they'd gone there to seek refuge, to seek shelter, because they were all over the place.

Some were together. Some were apart. It was a scene. We think there was some sort of environmental event. Something happened, and they'd gone to seek refuge in this cave. And while seeking refuge and hiding from whatever was outside, they'd died of starvation.[391]

Again, we're confronted by the specter of wanting to believe these are superior beings, and yet the evidence suggests that they may be understandably human in some of their motivations, and yet in others, completely alien.

De Comarmond continued:

We feel this was a genetic experiment. We're being told that there are probably hundreds or thousands of these beings buried in other caves in the region. There wasn't just one lone cave with these beings. There's more. What I think happened is that the extraterrestrials visited our home and went to a very remote place like Peru, where they could move freely because it wasn't heavily populated. And the Peruvian people have it in their belief system that we were seeded by extraterrestrials.

It was probably a situation where they could feel comfortable walking amongst humans as well, if there were any around. The area around Nazca is very remote. With all of the genetic experimentation that extraterrestrials seem to be doing, it seems they were looking at ways of combining their DNA with our DNA to find a way to allow them to live on our planet.

We're humans, and we like to think we are the almighty species and we rule the earth, and all the other animals are just weird little insignificant creatures who don't mean anything. But when the extraterrestrials came here and saw us, maybe we were leveled down to being just another species. Do you get that? They didn't buy into our hierarchy. And so, they probably thought, "Okay, well let's try it with a human. Let's try it with that monkey which looks dexterous and interesting."

I think there have been multiple experiments that have been done, and not just the ones we've seen so far.[392]

Heckenlively asked her about what it was like to see the ones described as insectoids, and she talked about the experience with Michael and the camera crew on location.

We're not really into the witchy, woo-woo kind of stuff, but we do believe in the endless possibilities of the universe. When we saw these beings, these insectoids, it was kind of a hive feeling we all had. They looked so fragile, and we all had the sense of wanting to protect them. It was hard to explain. Imagine you had a butterfly, but it was so big, you could see its features, and it could smile at you, and you could connect with it. It was kind of like that—this beautiful insect, huge and brought to life.[393]

When Heckenlively asked her about concerns that there were so many different types of bodies, from the insectoids to giants, and whether there might be some trickery involved, she replied, "To me, they're either all real, or it's all bullshit."

Heckenlively then asked if there was anything she thought important for the book that he hadn't yet asked. She related a story of a visit she'd made to Teotihuacán, an ancient pyramid complex in southern Mexico.

We met with the historian, and she walked us around and inside the pyramids and told us their history. The story the local inhabitants tell is that beings from a planet that was red came down to Earth, because the natural resources on their planet had been exhausted. They weren't coming to pillage here, but to live. They'd use Earth as a pitstop while they regrouped and recovered from the death of their planet, before they'd move on to somewhere else.

When they came to Mexico, they chose a random spot to land, and that was Teotihuacán. They terraformed the area around the site, changing the color of the soil to red using iron oxide. The claim is, they wanted the soil to resemble that of their home planet. It makes you think of Mars, right? Everywhere you go around the site, the soil is red. It's really weird.

The second thing she explained is all the hieroglyphics on the pyramids and the purpose of the pyramids. The hieroglyphics depict these beings. They actually have carvings of these beings on the pyramids. So, you find these beings depicted in Mexico, and in Peru at the Nazca lines. All of these ancient sites depict these beings.

So, definitely something has gone on at these places. Whether it's extraterrestrial or human experiments, who knows? But what I know is that these beings were alive, they lived on this planet, and they died here. And as a person who's had the opportunity to see them and touch them, I have a responsibility to figure out where they came from and what they were.[394]

Heckenlively spent some more time talking with Serena about his concerns regarding aliens, their abductions of human beings, and their apparent tinkering with our genetic code, but she didn't share his fears. Instead, she believed that the growing movement for disclosure was a cause for enormous optimism. She said:

I think we're in for a wild ride. Here's what I deeply believe. Human beings, you and me, we walk around and we've got family and friends. Maybe we've got a dog, and we have work that gives our lives meaning. Maybe we write books that influence people, and we should feel pretty good about ourselves. We should be pretty happy.

But all of us walk around with this hole inside us. The feeling that we don't fit in. There's always this uneasiness, this feeling we can't shake, no matter how great our life might be. We're like orphans who don't know where we came from. We don't know our purpose.

And once we have that knowing, once we can answer those questions, I think humanity's joy, our happiness level, will rise in such a significant way as to change our world forever. Imagine the world where we were able take away that sadness and fear from the common person.

It won't be about the tech; it won't be about curing cancer—although I'm sure that will happen. It will be about the feeling, the knowledge, that we're not alone. We have purpose; we have meaning. We're here for a reason, and it's an important one."[395]

The authors can't help imagining our world after we've answered, finally, that big question: are we alone in the universe?

Does humanity come together, or does it fall apart?

▲ ▲ ▲

In May of 2024, an article was published in the Spanish language journal, *Environmental and Social Management Journal* (*Revista de Gestão Social e Ambiental*, in Spanish), regarding an examination of one of the Nazca mummies known as "Maria." The results and discussion section reported:

> The tomographic imaging analysis showed that the specimen is a desiccated humanoid body with a biological architecture similar to that of a human, but with many morphological and anatomical structural differences such as the lack of hair and ears, an elongated skull and an increase in cranial volume (30% greater than humans); maxillary and mandibular protrusions as well as protrusions of the eyeballs, absence of the fifth lumbar vertebra, tridactyly in both hands and feet, in addition to different foci of arthropathies. Carbon-14 dating analysis of the specimen gave an average age of 1771 +/- 30 years, corresponding to 240 AD-383 AD. (after Christ). Implications of the research: It is demonstrated with further studies that this is a new humanoid species, it would have a strong impact on biology and science and scientific-historical and socio-cultural implications.[396]

This scientific validation of the genuineness of the specimen is an enormous step forward in solving the mystery of the Nazca mummies. It's often said that "extraordinary claims require extraordinary evidence" and the Nazca mummies may fulfill that requirement. Further independent testing is required to fully validate this claim.

The authors believe the mummies are probably the strongest evidence we have to date of nonhuman, advanced intelligences operating on our planet.

Michael reports that Peruvian officials are having private conversations among themselves for an official declaration of the authenticity of these specimens, but no final decision has been made.

⅄ ⅄ ⅄

On June 5, 2025, Michael Mazzola invited Congressman Eric Burlison to Mexico City for a remarkable series of events. As Michael explained in a text message with the Congressman and Kent Heckenlively:

All the witnesses and scientists involved in the Buga Sphere will be gathering in Mexico City and then on the 20th will be participating in a press conference at the university to present all of their evidence. I've seen this thing in person and some of the preliminary tests are incredible. If you are able to come to Mexico City during this period I will also arrange a private showing of the Nazca "Alien" mummies (which are absolutely extraordinary).[397]

The timing was excellent for Congressman Burlison, as his wife was at a work conference and his daughters were at a church camp, essentially leaving him at home as a bachelor.

Why not take a few days to go to Mexico to investigate alien artifacts?

The Buga Sphere was a new artifact in the UFO debate, and the history of it was recounted in a *New York Post* article from May 25, 2025.

A mysterious metallic sphere that was recovered after flying through the air in Colombia has left scientists baffled, with many speculating it's a UFO.

The bizarre sphere was recorded flying over the town of Buga in the western part of the country in March before landing and being confiscated, according to X user Truthpolex.

The orb-which reportedly weighed about 4.5 pounds and was cold to the touch when found-had etched into it a series of ancient-looking symbols, including runes and characters from the Ogham and Mesopotamian writing systems.[398]

Burlison asked David Grusch (who at the time was working for the congressman's office as a "special government employee), whether the congressman should consider the trip. Grusch replied that he didn't have an opinion about the Buga Sphere, and that while be believed some of the mummies were fake, some were "so intricate" and had matched up "with some of the things he had seen in a secure setting."[399]

The Buga Sphere was confusing to Burlison as it seemed to be both futuristic and poorly constructed at the same time. As he recalled, "They started talking to me about Tesla coils and how they thought it might work. And they talked about the equator that goes around it and the different holes. And when I was looking at the holes, I thought, 'These holes are not even properly aligned.' And the sphere is not exactly a sphere. It's oblong. And the writing on the sphere is not precise. It just seems absolutely like it's handmade."[400]

The only scenario which made sense was some sort of lost time-traveler, like in Mark Twain's famous book, *A Connecticut Yankee in King Arthur's Court*. Burlison saw some of the smaller alien figures which had previously been displayed before the Mexican Congress, but did not find them especially convincing. However, Burlison was allowed to examine and hold a large tridactyl hand and reported that it felt "very foreign to me."[401]

Burlison was impressed by the scientists who accompanied Dr. Greer, a group which included some former NASA scientists. Eric said, "They're testing the bones to determine the DNA of the mummy. They also took a chunk of the metal off the exterior of the sphere, and they specifically chose the side of the sphere that they claimed had hit the power line. The people who own the sphere were okay with them carving out one gram of material from the sphere. And then they carved out one of the numerous optical holes that are around the equator. They think they can carbon date that. They said they can ascertain whether the metal was created on Earth, or is derived from Earth, or from another planet. I have no idea how that science works. But that's what they were doing, and I believe these scientists."[402]

Until we finally have definitive results, which have been subjected to rigorous analysis, the situation will probably remain the same. Some will continue to see a hoax, a Fiji Mermaid, while others will see something truly remarkable, an interplanetary platypus perhaps.

If these tridactyl beings are real, and not some cheap taxidermist's trick, a great deal of our history will need to be rewritten.

CHAPTER TWELVE

WHAT IS REASONABLE TO BELIEVE?

In putting this book together, we've tried to fairly present both sides of the issue, detailing the claims of those who believe in UFOs and aliens while allowing the self-proclaimed skeptics, like Congressman Eric Burlison, to have their full say as well as presenting the findings of the March 2024 All-domain Anomaly Resolution Office (AARO) report—which discounts the possibility of alien life visiting Earth—and other congressional inquiries.

However, we've also weighed in with our thoughts on the strengths, weaknesses, or contradictions of certain claims by various individuals.

A commonality between both believers and skeptics appears to be that the government is hiding something. Whether what it's hiding is an entirely human-developed advanced technology program or something at least partially based on recovered alien technology and biological material is less clear.

Another observation we believe is fair to make is that both skeptics and believers have concluded that the government whistleblowers are reporting their information in good faith, even if they may be mistaken about the origin of the vehicles or phenomenon they've observed.

Even at this stage, we're uncomfortable suggesting to anybody we know the ultimate answer to the question of whether we have been visited by aliens. We're intrigued by the claims of many and convinced of the sincerity of the witnesses and want to know more.

We need more—possibly better DNA testing of the Peruvian mummies, and a great deal more disclosure from the United States government. And if we're to believe anything, we need more genuine engagement in the scientific community.

However, if it's true that alien beings have been manipulating human DNA for a long time, it may answer one of the essential questions about our own development, if not our origins.

ᴧ ᴧ ᴧ

What has puzzled anthropologists for decades is that we are much smarter than our primate relatives, and this development seems to have taken place fairly recently.

An article in *Stanford* magazine addresses this mystery through the eyes of anthropologist Richard Klein, a University of Chicago–trained researcher and acknowledged thought leader in his field. The article sketches out the conflict in the following manner:

> Most—though not all—anthropologists agree that human culture, imagination and ingenuity suddenly flowered around 45,000 years ago. The evidence ranges from fantastic cave paintings and elaborate graves to the first fishing equipment and sturdy huts. And whether scientists call it the great leap forward, the dawn of culture or civilization's big bang, they agree the change was momentous, giving humans the cohesion and adaptability to expand their range into Europe, Asia, and eventually Australia and the Americas. "In its wake," Klein says, "humanity was transformed from a relatively rare and insignificant large mammal to something like a geologic force."[403]

In light of the DNA results from the Peruvian alien mummies, the question is why human beings made such a great leap forward around 45,000 years ago and whether there might have been some outside influence. The article continues with this mystery of our prehistory:

> To witness the contrast between premodern and modern ways of life, Klein says, sift through the remains from caves along the southern coast of South Africa. Simple Stone Age hunter-gatherers began camping here around 120,000 years ago and stayed on until around 60,000 years ago, when a punishing drought made the region uninhabitable. They developed a useful tool kit featuring carefully chipped knives, choppers, axes and other stone implements. Animal bones from the caves show that they hunted large mammals like an Eland, a horse-sized antelope. They built fires and buried their dead. These people, along with the Neanderthals then haunting the caves of

Europe, were the most technologically adept human beings of their time.[404]

In other words, physically modern humans were around for tens of thousands of years before they started using their brains the way we do today. It's generally acknowledged that human brains reached their current size around 130,000 years ago, which leaves a gap of approximately 85,000 years before we started displaying the creativity and talents typically associated with modern humans. The article continues:

> However, Klein says, there were just as many things they couldn't manage, despite their modern-looking bodies and big brains. They didn't build durable shelters. They almost never hunted dangerous but meaty prey like buffalo, preferring the more docile Eland. Fishing was beyond their ken. They rarely crafted tools of bone, and they lacked cultural diversity. Perhaps most important, they left no indisputable signs of art or other symbolic thought.
>
> Later inhabitants of the same caves, who moved in around 20,000 years ago, displayed all these talents and more.
>
> What happened in between?[405]

The fossil record would seem to indicate a specific time and location for a dramatic change in human behavior, without any accompanying change to humans' physical appearance. While many argue that the change was gradual, or that there was a slow buildup and a sudden explosion of creativity, not leaving evidence in the fossil record until roughly 45,000 years ago, Klein suggests that the fossil evidence is accurate.

> Klein suggests a third possibility—a strictly neurological scenario that has gained few followers in a field of study dominated by cultural explanations, he says. Humanity's big bang, he speculates, was sparked not by an increase in brain size but by a sudden increase in brain *quality*. Klein thinks a fortuitous genetic mutation may have somehow reorganized the brain around 45,000 years ago, boosting the capacity to innovate. "It's possible this change produced the modern ability for spoken language," he says.

Clearly, speech eases communication. But it also fosters something less obvious and equally important. Spoken language, Klein says, "allows people to conceive and model complex natural and social circumstances entirely within their minds."[406]

A "fortuitous genetic mutation" 45,000 years ago turned us into modern humans? It's a convenient scientific explanation to a question in the fossil record for which we have no genuine answer.

And since we have no direct evidence of what caused this change, all we can do is speculate. Perhaps the genetic mutation that gave rise to an entirely new set of capabilities in humanity was a fluke. But if we are to consider the information from the Nazca alien mummies, we must at least entertain the notion that an alien being could have tampered with the DNA of our species, setting us on the road to civilization and our modern world.

Your very ability to consider the ideas in this book may be the result of alien manipulation of human DNA in the distant past.

Suddenly, the UFO issue isn't of importance just to those who see strange craft in the skies, claim that there have been crashes, or report encounters with alien beings. It's of importance to every person on this planet.

The origin and development of humanity itself may have been impacted by some form of alien intervention. Is that the information the guardians of the UFO secrets don't want us to know?

Is this the "catastrophic disclosure" they are trying so desperately to avoid?

ᛉ   ᛉ   ᛉ

Although many authors have suggested that aliens have assisted in human evolution, probably none is as well-known as author Zecharia Sitchin, whom journalist Corey Kilgannon profiled in a 2010 *The New York Times* article.

The old man knew how the mainstream press was likely to portray his work, but Kilgannon, to his credit, was remarkably fair-minded, at least in describing Sitchin.

> "Well, you could start by calling me the most controversial 89-year-old man in New York," Mr. Sitchin says. "Or you could say I write books. I understand you've got to have an opening sentence, but describing my theories in a sentence, or

even something like a newspaper article, is impossible. It will make me look silly."[407]

Kilgannon seems to have a certain fondness for the older man, not believing his theories but wanting to give credit to his thirteen books, which have sold millions of copies.

He slides over a cup of coffee with a 30th anniversary logo for "The Twelfth Planet," his seminal first book, now in its 45th printing. It stated his basic theory, based largely on his readings of texts preserved on clay tablets from the pre-Babylonian era in ancient Mesopotamia, the so-called cradle of the civilization of Sumer....

Starting in childhood, he studies ancient Hebrew, Akkadian and Sumerian, the language of the ancient Mesopotamians, who brought you geometry, astronomy, the chariot and the lunar calendar. And in the etchings of Sumerian pre-cuneiform script, the oldest examples of writing, are stories of creation and the cosmos that most consider myth and allegory, but that Mr. Sitchin takes literally.[408]

The article also notes that Sitchin had studied economics in London, worked for most of his life as an executive at a shipping company, had been married for sixty-six years, and raised two daughters, all while spending his free time studying and leading tour groups to archeological sites, where he explained his controversial theories to his fellow travelers.

Sitchin was not just a normal man but an exemplary one, according to Kilgannon. And to the journalist's credit, within the limited space of a newspaper article, Sitchin was allowed to give a brief overview of his theory.

It starts with the planet Nibiru, whose long, elliptical orbit brings it near Earth once every 3,600 years or so. The planet's inhabitants [Anunnaki, the name given them by the Sumerians] were technologically advanced humanlike beings, Mr. Sitchin said, standing about nine feet tall. Some 450,000 years ago, they detected reserves of gold in southeast Africa and made a colonial expedition to Earth, splashing down in what is now the Persian Gulf.

Mr. Sitchin said these Nibiru-ites [Sitchin's name for them] recruited laborers from Earth's erect primates to build eight great cities. Enki, who became the Sumerians' god of science, bestowed some of the Nibiru-ites' advanced genetic makeup upon these bipeds so they could work as miners.[409]

Were Earth's erect primates simply waiting around for an alien race to appear, alter their genes, and put them to work as miners? Maybe these aliens had mind control technology to make the daily lives of their primate slaves bearable, unlike the barbarity of human slavery. But at its core, it was still slavery. The article concludes:

This is how Mr. Sitchin explains what scientists attribute to evolution. He says the aliens' cities were washed away in a great flood 30,000 years ago, after which they began passing on their knowledge to humans. He showed a photograph of a wood-carving from 7,000 B.C. of a large man handing over a plow to a smaller man: Ah, the passing of agricultural knowledge. Anyway, he said, the Nibiru-ites finally jetted home in their spacecraft, around 550 B.C.

"This is in the texts; I'm not making it up," Mr. Sitchin said, finishing his coffee. "They wanted to create primitive workers from the homo and give him the genes to allow him to think and use tools."[410]

To add a little more context to Sitchin's theories, the Anunnaki were mining gold on Earth as a means of protecting their own atmosphere, which had apparently sustained a great deal of damage.

We're certainly not suggesting that Sitchin's beliefs are correct. (In fact, one claim that seems completely mistaken is that in 2016, the alien home world of Nibiru would become visible to us.) But one must note the similarities between Sitchin's account of what may have happened and the accounts of other ancient stories, such as regarding the Nephilim in the Old Testament Book of Enoch, which describes them as large, muscular giants who married human women. This is what Genesis says:

When human beings began to increase in number on the earth and daughters were born to them, the sons of God saw that the daughters of humans were beautiful, and they married any of them that they chose. Then the LORD said, "My Spirit will not contend with humans forever, for they are mortal; their days will be one hundred and twenty years." The Nephilim were on the earth in those days—and also afterward—when the sons of God went to the daughters of humans and had children by them. They were the heroes of old, men of renown.[411]

We have to admit that the Bible's account of the Nephilim raises just as many questions as Sitchin's claims about the Sumerian Anunnaki. Is it simply part of human nature to imagine beings from the heavens, whether they be angels or aliens? Or is it the preserved history of events in the distant past, as well as modern accounts?

Even earlier than Zecharia Sitchin's claims was Erich von Däniken's 1970 book *Chariots of the Gods?*, which first raised the possibility of ancient visitors to our planet. Fifty-five years later, the book's introduction raises some of the same questions many have been raising today.

Nevertheless, one thing is certain. There is something inconsistent about our past, that past that lies thousands and millions of years behind us. The past teemed with unknown gods who visited the primeval earth in manned spaceships. Incredible technical achievements existed in the past. There is a mass of know-how that we have only partially rediscovered today....

I claim that our forefathers received visits from the universe in the remote past, even though I do not yet know where or from which planet they came. I nevertheless proclaim that these "strangers" annihilated part of mankind existing at the time and produced a new, perhaps the first *homo sapiens*.[412]

None of us knows the answers to the many questions posed in this book. Do aliens exist? If so, what are their intentions towards us, and what is our true history with them?

Zecharia Sitchin does not know.

Erich von Däniken does not know.

Nor do any of the people we interviewed for this book, regardless of how strongly they may hold their beliefs.

But many are asking the right questions and demanding that what is known, be revealed to the general public.

Catastrophic disclosure can mean many things to many people.

For some, it can mean we did not naturally evolve into what we are today, thus destroying their faith in evolution. We did not emerge from nature as the superior species on the planet but may have been engineered for purposes having nothing to do with our welfare and benefit, by creators who were, at best, indifferent to our continued survival.

For others, it can mean we were not divinely created by an all-powerful God, thus destroying their faith in religion. We may have interacted with beings from the sky, but they were not angels or messengers from God; they were visitors from the stars.

Catastrophic disclosure might mean to others that for more than eighty years, we've been lied to about what has been happening in our skies, a deception perpetrated by governments around the world.

For Serena DC, it means that we will finally stop feeling alone, as if we're the orphans of the universe, and take our place among our rightful family.

For Michael Mazzola, catastrophic disclosure means something completely different. For him, the real story isn't about the aliens but about the humans. The ones in the shadows who've controlled the power of free energy and unbelievable technology that can not only raise people out of crushing poverty and cure our diseases, but promote a more peaceful world, where we all have what we need. The catastrophic disclosure has nothing to do with the aliens, but with the all-too-human villains who have kept us from living in a world of prosperity. Nobody needs to live on a poor, polluted planet, where so many suffer from disease. We can make our world a paradise.

For Kent Heckenlively, he just wants to know what's true. He'll figure out all the implications later.

⅄   ⅄   ⅄

Let's propose at least a partial narrative that makes sense and seems to fit all the claimed facts.

The explosion of the first atomic bomb on July 16, 1945, broadcast to somebody not on Earth that we humans had reached a new and potentially dangerous stage in our evolution.

The first craft to respond was poorly prepared, and crashed about a month after July 16, resulting in the chain of events reported in the 2021 book *Trinity: The Best-Kept Secret*, by Jacques Vallée and Paola Leopizzi Harris. The crashed vehicle was first reported by two children on horseback, and it seems as if the military struggled with crash recovery, suggesting that this was their first experience with a crashed vehicle. The two children also reported that the beings they saw struggling to escape seemed to have some telepathic ability, by which they expressed that they were in pain.

In the summer of 1947, the aliens made another attempt to get close to the nuclear test site, but in Washington state they had trouble over Maury Island and were seen by pilot Kenneth Arnold near Mount Rainier. Then, shortly after that, there was another alien crash in Roswell, New Mexico. But this time the military had a better-trained recovery crew and a media cover story ready.

From that point, it appears that the US government, under the cover of Majestic-12, took charge of the alien investigation, publicly casting doubt on the story while also investigating the issue and trying to reverse engineer the discovered technology.

The name Wernher von Braun is often associated with these efforts, and his involvement seems to be something of an open secret among many members of the scientific community, such as astronaut Gordon Cooper.

In 1964, with the abduction of Betty and Barney Hill, the alien abduction narrative took off, and there were troubling reports of the aliens' taking sperm and eggs from the abductees for unspecified reasons. The majority of the UFO community seemed to eventually accept this phenomenon as real, but a smaller group of individuals, perhaps best exemplified by Steven Greer, suspect that these were staged abductions by US military personnel, using reverse engineered craft and technologies.

The public, including Presidents Carter and Reagan, continued to see UFOs. Senator Barry Goldwater, the 1964 GOP presidential nominee, chased a UFO in his single-fighter Arizona National Guard jet.

The 1980s were dominated by President Reagan's Strategic Defense Initiative (popularly known as "Star Wars"), which sought to develop technologies that could shoot down nuclear missiles in flight. Some suspect this was a cover for developing advanced weaponry against alien visitors.

In the 1990s, President Clinton made several approaches to the intelligence agencies to learn about UFOs and was consistently denied access to this information.

The first decades of the twenty-first century were defined by the Global War on Terror, launched in response to the September 11, 2001, terrorist attacks on New York City and Washington, DC, and by the development of a massive surveillance state. Did that give us additional technology with which to monitor the "visitors"?

Starting in 2017, with the publication in *The New York Times* of credible claims by Navy pilots about UFO encounters, public demand for some type of disclosure program grew.

In 2023, members of the military, released from their security oaths, testified in front of the US Congress about their UFO experiences and about highly classified government programs that might be using reverse engineered alien technology.

Also in 2023, there was a presentation in front of the Mexican Congress purporting to show 1,700-year-old Peruvian mummies with three toes and three fingers, who appear to be genetic hybrids using the DNA of humans, chimpanzees, and something else. That effort in Mexico and Peru continues.

When we consider this narrative, a few preliminary conclusions come to mind.

The first is that it must take a great deal of energy and resources to visit our planet, because it does not appear that aliens are here in large numbers.

The second is that the aliens can change their technology in a reasonable time frame, as shown by the fact that most of the reported UFO crashes and recoveries took place in the twenty years following the first atomic explosion.

The third is that the type of being most commonly encountered, the grays, doesn't seem to genuinely represent the aliens; rather, these beings have been bred to more effectively work on our planet. Instead of weapons, it seems they have powerful mind control (telepathic) attributes or technology, which allows them to move safely among the local inhabitants, who are most likely shocked into some form of paralysis, unable to fight back or escape.

The fourth is that the aliens may have a morality far different from our own, one that justifies large-scale genetic tampering, which the Nazca mummies could indicate.

The fifth is that the aliens may have tampered with human DNA in the distant past, setting us on the road to civilization, while at the same time using us for some amount of time as slave labor.

We believe that if the aliens were to come out of hiding and show themselves, humanity would do a lot of maturing, likely putting war and a great deal of human mischief aside, as we saw new worlds of wonder beckoning to us.

By the same token, we don't believe that those in the government who have kept these secrets, and who have attacked the witnesses and whistleblowers, have done much to recommend themselves, either.

Are they keeping secret technologies that could change our world, as Steven Greer believes?

We know the truth is out there.

What if the truth is that we are, in essence, the latch-key kids of the galaxy, conceived not in love by a supreme being but because an alien species needed us for some purpose, and then abandoned us to our own devices, with no known brothers or sisters among the stars?

Can we still find God in such a picture?

Science provides a different narrative, one of competition between various hominids over millions of years—a competition that refined human beings until they stood at the very pinnacle of creation. Would the reality of alien intervention in human development destroy our faith in both God and science?

Do we stand as lords of Earth because we've conquered all challengers, or is it more accurate to view ourselves as slaves whose masters have fled the plantation for reasons unknown? We need stories to give our lives purpose, and the idea of aliens' taking a hand in human development destroys, or at least alters, the stories we have told ourselves to date.

When we look at the horrors of human history, might we blame at least some of it on the aliens—who, after giving us intelligence, quickly took us into bondage and slavery? Are we brutal because that is our nature, or is that the patrimony these ancient aliens have bequeathed to us?

Is the fault in ourselves, or because we are the children of oppressive parents?

If we consider ourselves, at least in part, "children of the stars," perhaps we might take some solace in the fact that children are often radically different than their parents, each having their own unique destiny?

Maybe that is where God, science, karma, or whatever you might prefer to call it comes into play.

In order to become healthy adults, we must accept those things about our past that we cannot change, the influences from our parents we do not like, but must also acknowledge that the ultimate direction of our lives and accomplishments is uniquely our own responsibility.

Perhaps that is what we must eventually do as a species in relation to any possible alien influence on our development.

Only by understanding the past can we find our future.

We believe that if we make the right choices today, if we consider our history without passion or prejudice, the years ahead can find humanity in the stars; and we can move forward with genuine knowledge of potential allies and adversaries as we embark on our next bold adventure.

STEVEN GREER WEIGHS IN ON DISCLOSURE EFFORTS AT START OF THE SECOND TRUMP ADMINISTRATION

Perhaps it's fitting that this book ends where it began, discussing the issue with Steven Greer, who for more than thirty years has made disclosure of the truth about UFOs his life's mission.

Michael Mazzola has a podcast with Greer, on which the two of them discuss various aspects of the UFO phenomenon. On February 13, 2025, they recorded a long podcast titled "Catastrophic Disclosure,"[413] in which they discussed the issue at length.

In Heckenlively's initial discussions with Greer, he was skeptical of Kent's ability to fully contemplate the enormity of the issue, and the level of deception perpetrated by those in power. We've tried to keep that concern uppermost in our minds as we've worked on this book.

And yet, despite his warnings, we find the narrative he proposes to be simple to comprehend: in the absence of genuine transparency, human beings will perpetrate the most wicked acts imaginable, and probably a couple more that decent people would never even contemplate.

Michael begins that episode of the podcast by asking, "What is catastrophic disclosure, and for whom is it catastrophic?"

Greer responds:

> The term was first used by some Special Forces guys who were moving forward in the last two years to come forward with a great deal of this kind of information. Evidence of crash retrieval and all of that. They were told by some people in the intelligence community, "Well, that would be catastrophic disclosure to do that." And I actually think that's a type of gaslighting of whistleblowers. But there is a type of catastrophic disclo-

sure of which I am afraid. And that's if this comes forward in a way that's been planned for decades, which would be much worse than COVID, much worse than 9/11.

More people believe that we've been visited and that the government's been hiding things for decades than have voted for any president in modern times. It's up in the 60 percent-plus range. That's bigger than the Reagan landslide, and even very few members of Congress have those kinds of numbers. I think you have people embedded in government who keep trying to put speed bumps in the way of this information coming out, especially the technologies. They keep saying, "Oh, it's just too much, and the people can't handle it." That's not true. The people would love it.

Here's who can't handle it: the permanent bureaucracy of corrupt interests, both in the military-industrial complex and, more importantly, the financial complex. The technologies, the energy, the zero-point quantum vacuum technologies can't come out, because you've got trillions of dollars invested in commodities and oil. It should have come out a hundred years ago. That's why we did our film, *The Lost Century*. What is really catastrophic is this information not coming out, or coming out in a way that's been planned for seventy years, since the 1950s, that this is an existential threat to humanity. There are these forces out there, and they're calling them NHI [nonhuman intelligences], which of course is an intelligence term, a concocted cover-up term to present them as a threat to the human species.

Catastrophic disclosure is not acknowledging that we're not alone in the universe. It's acknowledging that they've been here, we've downed their craft, reverse engineered them, or even that we have ones of our own that we've built. That would be what most people are already beginning to say, even in Congress. The catastrophic part of it, and this is what's going on, is that whistleblowers, members of Congress, even people

in the new administration, are being gaslit by very slick counterintelligence operatives who present this as a threat.[414]

In the interview, Greer went on to warn again about the efforts of Luis
Elizondo, whom he believes to be one of those "very slick counterintelligence operatives" who is getting the public to look in one direction while
keeping them away from the truly disturbing question of what governments
around the world have done, not only with the reverse engineering of these
craft but with the operation of these craft and their technologies, and how
governments have used them to control their own populations. What seems
to genuinely enrage Greer is when people like Elizondo go on large pod-
casts, such as *The Shawn Ryan Show*, and claim that these objects are real,
and must therefore be alien, because no nation on Earth has such technology in its arsenal. Greer says of Elizondo's statement:

> Totally a lie. Now, why would that lie be so valuable? Think
> about it for a minute. If you go to the Disclosure Project
> Intelligence Archive, dpiarchive.com, you'll see 120-plus crash
> retrievals. You'll see witnesses, military and corporate wit-
> nesses, to these man-made UFO. People like the recent whis-
> tleblower, Jake Barber, have repeatedly said to us that they [our
> own craft] are indistinguishable from the actual ET craft. The
> technology is that advanced in what Lockheed Skunk Works,
> Northrop Grumman, Raytheon, and a few other big contrac-
> tors have that build these things. We also know they're being
> built in Russia, and most likely China. By the same consortium.
> This is a global criminal enterprise.
>
> I think the catastrophic disclosure is the secrecy that hides the
> technologies that would benefit the earth, because they keep
> saying, "Well, we don't know how they operate. We don't know
> how zero-point energy is. Nothing like that has ever been
> built." Total, fabricated, bold-faced lies.
>
> That leaves us dependent as a civilization on an 1800s energy
> paradigm for our energy system. This not only requires most of
> the population of the world to be impoverished to some degree,
> but also under control of a centralized system. It's actually what
> I call petro slavery, where the whole planet is dependent on fos-

sil fuels and the centralized power grid for their very existence as a modern society.

That would all change if these technologies came out. It would be the ultimate liberation of people in all strata of society, in the United States, the developed world, and also in the developing world. All of us would be liberated. Who is that catastrophic for?[415]

The coauthors have discussed Greer's point several times during the writing of this book. The issue is not really about aliens or what type of craft they might zip through our skies and oceans (fascinating as that subject might be), but what this information might mean to every person on Earth.

Forget every argument you've ever had about politics or mankind's effect on the environment. Since the dawn of civilization, humans have lived in a world of apparent scarcity, and if we wanted our tribe or nation to survive, we needed access to certain resources, usually held by other groups. We would either negotiate, trade, or go to war with such groups (and often a combination of the three) in order to acquire those resources.

But what if we flipped the table on that game and said, "Here is abundant, nearly free energy, and you will have what you need to produce food and water, clean up the environment, travel the cosmos, and live a life free of debilitating diseases."

However, those in the shadows, those who have both the alien technology and the reverse engineered craft, do not want to give up their precious power, and are willing to go to great lengths to maintain their current positions. Greer continued on the podcast, detailing how this secrecy is being maintained:

Withholding this information and saying we don't have these technologies is itself a type of continuation of human global enslavement to the current system. Let me add one other thing. These people have portfolios that they show people in national security circles. Let's say you're the president or you're on the Senate Intelligence Committee. These folks will come in with a portfolio of alleged NHI craft, or NHI beings, doing things to humans: abductions, mutilations, blowing up villages, vivisections, cutting people open alive, feeding babies to spider-like aliens. Right out of a horror movie.

Now, aside from the fact that a lot of it is made up out of whole cloth, there are parts that are real. There are real victims of abduction. I was almost abducted in 1994. Those are being done by humans masquerading as aliens. They're using this technology to do it. They're using craft that look like an alien ship. They're using technologies that are very advanced electronics, like supercharged Havana syndrome–type energy weapons that affect consciousness and all sorts of things.

These are being used by people who are masquerading as aliens and convincing officials there's this threat. But that falls apart if there's honest disclosure.[416]

We accepted the idea that powerful individuals and organizations would act in such a heinous manner much more quickly than we'd like to admit. The authors believe there's a specific portion of humanity who are sociopaths, who would willingly lead humanity on a path to destruction and enjoy the fireworks.

Greer goes on to detail what he believes are three groups at work: the aliens and their craft, humans masquerading as aliens, and interdimensional beings brought into our world through very advanced electromagnetic systems, similar to those seen on the Netflix series *Stranger Things*. The denial by these counterintelligence operatives that we have any equivalent technology drives the public to the mistaken conclusion that all these reports—the sightings, the cattle mutilations, the abductions—must be related to something nonhuman.

Greer continues:

It's about uniting the world under a militaristic sort of global totalitarian system where everyone is terrified of what's out there in space. And that's how you get people to give up their freedoms. As Benjamin Franklin said, "Those who would give up their liberty for security shall have neither." These people would like to control the populace through fear. That's the way the demagogues have done it for millennia, whether they be religious, political, or military leaders. It's the ultimate fear play.

And that's how you colonize people's minds. And that's how you end up in a catastrophic disclosure. It's that there's not enough of it in the right context. And this has been going on for the thirty-five years I've been doing this. Let's not forget that the Clinton administration was very close to disclosing all of this.

They did form a team, an operation called Team Red, to penetrate those programs and get them back under control. And Team Red was completely decimated—either killed or absorbed into the illegal black projects. And I think this is something we have to be very careful about. Because I believe we're at a point where everything is poised to come out.

There's growing bipartisan support in Congress, as you saw the other day with Congresswoman Luna, whom I have personally briefed on this issue. There are others, like Congressman Burchett and Moskowitz, [with] whom I've also discussed this issue. I think there's strong bipartisan support to get to the bottom of this. The problem is, they're being advised very badly by people who swoop in and create these false narratives of threat. And false narratives of "We don't have anything like this."[417]

Greer went on to discuss recent whistleblower David Fravor, the Navy commander and F-18 pilot who chased a UFO and filmed it (the Tic Tac incident off the coast of San Diego in 2014), and whose testimony was among those that opened this book.

When I first spoke to him [Fravor] a few years ago, he was adamant that what he had pursued absolutely had to be alien. That there's no way any human on Earth had any technology like that. And that would be a very rational assumption for someone having their first encounter with a propulsion and energy system like what Lockheed Skunk Works or Northrop Grumman or Raytheon has, because they do not move in an aerodynamic form. There's no rockets. There's no jets. The infrared sensors which would capture a heat signature? Nothing. So, what is making this thing move like that?

And it would be a normal assumption for a conventional military or a civilian pilot or a rocket scientist, for example, dealing with normal propulsion to say, "Well, that has to be from outer space or another nonhuman civilization."

And that is where people make the big mistake. Now, I talked with David Fravor a few years ago and said what he saw was from Lockheed Skunk Works. He has since admitted that someone from Lockheed has confirmed to him directly that this was one of theirs. So, that is a very good evolution, but it takes a while.

The problem is, I don't know if we have time for people to spend several years wrapping their minds around this, because I think we're on a fast track as to where this is all going to come out. And if our president and national security team and Congress, as well as leaders around the world, really think all of this phenomenon has to be alien or nonhuman, then that opens the door for exactly what the seventy-year evil agenda has been. And that is to have a threat that the whole of humanity faces.

To me, catastrophic disclosure is a false disclosure, done in bits and pieces but in a manipulated way. In a carefully orchestrated way that ends up driving policy in the 180-degree opposite direction it needs to go. But without enough information and enough disclosure of the facts, you could come to a mistaken conclusion like David Fravor did. It was an error, but a totally rational conclusion.[418]

In the interview, Greer goes on to note many of the recent positive developments in the political arena around the alien issue, not simply the recent reelection of President Trump but the team he has assembled around him. This includes people like Tulsi Gabbard, the new director of national intelligence, as well as Amaryllis Fox Kennedy, the campaign manager for the independent presidential candidacy of Robert F. Kennedy Jr., a former CIA officer, and now a member of a newly established intelligence advisory board for the president.

As previously mentioned, Michael Mazzola directed a film with Greer titled *The Lost Century*, which premiered in 2023. In it, Greer details the peaceful, nonpolluting world we might inhabit if alien technologies were shared with humanity. He returned to that theme again in this interview with Michael, speaking of "the energy generation systems that would give us free energy for our homes, our cars, our factories, our businesses, that would grow the global economy. And one of the arguments to be made to the powerhouse military intelligence mainstream people here in the United States, as well as Congress and the president, is that it would be a tide which could lift all boats But it could be led by the United States."[419]

Michael had told Heckenlively that Greer was consulting with Special Forces veterans who wanted to take possession of the downed and reverse engineered craft and put them under the jurisdiction of our elected government. It sounded fantastical to Heckenlively at the time, but Greer talks openly about it with Michael in the interview:

> About a year and a half ago I had a gathering with, I think it was the top seven Special Forces guys in the United States. They were being pulled out of the military and put together a part of an operation. It's a highly classified Special Access Project that would be in charge of getting these illegally run programs and operations under proper control and supervision.

> When they asked me what's the most important evidence and technology to seize, I said, "By far, the manmade craft, the systems that are simulating fake extraterrestrial events, the manmade aliens, and all the technologies which allow for that deception to take place." That is priority number one I've given them.

> And the second one, a very close second or maybe even equal to it, would be liberating the captured ETs that are in underground facilities. These are actual nonhumans that we've captured and are kept essentially as prisoners. They're euphemistically called refugees. That's actually one of the terms an operator/contractor used. I said, "Well, I don't think you're a refugee if you've had your craft shot down, been captured, and thrown into an underground facility."[420]

As much as we thought this was going to be a story about unidentified flying objects and alien beings, we find we have more questions about the humans involved in this story. If there is an entire hidden group in our world preventing information and technologies from being disclosed, how does it operate? What's the psychology of its members? How do they live and love, and what do they believe that justifies their actions?

In the interview with Greer, Michael asked a question that got to the heart of this inquiry:

MICHAEL MAZZOLA: Dr. Greer, you mentioned that the Legacy group is deeply fractured right now. And I think most people imagine this as some sort of Spectre-like group, like in James Bond films. They're all sitting around the table; there's one leader and one vision. And they all go out and execute that vision with all of these tentacles, and these tentacles act in perfect concert with each other. But what I've learned from you is that there's many competing agendas within this Legacy group, and it's driven by different religious ideologies. Could you unpack that a little bit for us?"

STEVEN GREER: I learned of this in the '90s, after we'd done the first National Press Club Disclosure event in 2001. I had someone from the organization say, "This has really moved this forward, and I think there's about 70 percent of who think this should now come out." [Greer believes there are about two hundred to three hundred members of the international policy committee.] The problem is, the other 30 percent are ultra-hardliners, the more violent and psychopathic members, and they're very likely to kill their own people. Former CIA director William Colby—he'd been CIA director under Nixon and Ford—said he was going to get us a very sophisticated zero point of quantum vacuum energy system, as well as about fifty million dollars in seed money, because he wanted to open source it so the world would have free energy.

He—Colby—thought it was time to do it, and the week he was going to meet with a member of my board, they found him floating down south of DC in the Potomac. His best friend, a full-bird colonel, told me that was absolutely a wet work. This was a targeted killing or assassination. The extreme folks in that organization, here and abroad, are not shy about killing their own people. There's a certain

level of threat and intimidation that goes on. The simplistic conspiracy theories that people have are all wrong, because that's not how
it operates. And the strangest part of it is the folks who are at the
far-fanatical-religious end of it.

Let me put it this way. I have no beef with organized religion. But I'm
talking about the extremists who want to bring about Armageddon.
Like Prince Hans-Adam von Lichtenstein told me, "We have to have a
war with the aliens," or at least pretend to, so that Christ will return.
And Armageddon has to happen before Christ will return. As bizarre
as that sounds, this is a serious group within that organization, and
they are true believers.

It's almost like they're the extreme version of Al-Qaeda or the Taliban,
or some terrorist group. But it's the extreme religiosity as opposed to
spirituality. I've been at the Vatican and talked to very reasonable
people about this. And yet there are folks in that organization as well
that are drawn from different religious groups that have those same
beliefs. The only way forward for humanity is to destroy it, so that
we're saved in the Rapture, or something like that.[421]

Again and again, we keep coming back to the idea that this story isn't
about the aliens or their advanced technologies. It's about us. Human beings.

Are we capable of grasping the moment we're in (if in fact we're not
truly deluded) and making the right choices?

In the study of history, one is always intrigued by the character of individuals who made enormous changes in their society, especially when those
societies were full of injustice. There is a through line of genuinely excep-
tional individuals, from Jesus to Mahatma Gandhi, Martin Luther King Jr.,
the Dalai Lama, and Nelson Mandela. They were or are relentless warriors
against evil, but gentle to all people, including their enemies.

The attempt is to bring wayward souls back into the fold, rather than
destroy them. The effort is always to heal, even those most severely cor-
rupted by the enticements of darkness.

But until there is honesty and reconciliation, we're probably going to
need some well-trained soldiers of the type with whom Greer has been
meeting. The difficulty in any battle, though, is knowing the capabilities

of your enemy. It's one thing to be brave; it's another to be massively out-gunned by a superior level of weaponry.

Greer continued on this theme:

> Until there is some kind of well-trained, well-advised team to go in and stand down those systems, it probably ain't going to happen. And the other problem with that is that the technologies that the Legacy organization have are so much more advanced than a conventional law enforcement or military group. It's a bit like native people throwing spears into the rotors of a Sikorsky helicopter gunship. It's daunting. And this is one of the big problems.

> Now, it can be dealt with, and I'm advising some of the guys who are willing to risk their lives to do it, But we need an executive order that clearly authorizes the operation and funds it properly. Because I don't think it can be done for less than five hundred million to a billion dollars. And that's just getting started. Because we're not just talking continental United States facilities.

> We're talking the South Pole and Antarctica and South America and the Australian outback, all these places where there are underground and covert operations taking place. And we keep talking about this. Jacques Vallée had a CIA document from 1985 where they were talking about staging alien abductions in Brazil and Argentina for their psychological warfare value. This is the heart of the problem. It is catastrophic disclosure if nobody can tell the difference between a staged human operation and an actual extraterrestrial civilization.[422]

Michael continued his conversation with Greer for several more minutes, but they'd made their important points. The reverse engineering was getting so successful that the dark forces were masquerading as aliens to spread fear as a way for the tyrants to take control.

The villains would seek to make the public believe they were the heroes.

Heckenlively called Michael after listening to the podcast and asked if he could get Greer to talk to him one more time.

The interview was set for Wednesday, February 19, 2025.

▲   ▲   ▲

Heckenlively began the interview with Greer by reminding him that when they'd last talked in July of 2023, Greer expressed great skepticism over his ability to get the story right. Heckenlively told Greer he'd kept his warning uppermost in my mind over the past year and a half, and hoped he'd be pleased with the final product.

Heckenlively also told Greer he'd found Greer's narrative to be the most internally consistent, and noted he'd been saying the same thing for more than thirty years: the major problem is that when you create systems that lack transparency, you invite the very worst of human behavior.

Heckenlively began by asking him about his recent efforts with the new Trump team, coming up on a month into the president's second term.

Greer said:

> I've just come back from some meetings at Mar-a-Lago and elsewhere with some senior people in the administration who are just now getting up to speed. I'd rather not mention who at this point, but relevant national security–related folks. A couple things I already knew but have become verified.
>
> Number one, the White House has very little, if any, legitimate information on this subject. Very important. Number two, the people coming in, and that would include the president, are even less informed, unless they've already had a counterintelligence group attached to them. One person I met with, who is now a cabinet member but not related to national security, had been gaslit by Luis Elizondo and some other people.
>
> For example, that person had been told, number one, all of these [craft] are alien, because we don't have anything that flies like that. Of course, that's false. That is the narrative from the counterintelligence and disinformation community for the express purpose of being able to then run a false flag operation. It would all then be blamed on the "aliens" or "NHI crap."
>
> There is a big problem, because I know that Don[ald] Trump Jr. has been specifically targeted with that false information. We know people in the House Oversight Committee have been,

and we also know there are people in the upper echelons, cabinet level, of the Trump administration who are innocent victims. I'm not saying they're part of any cover-up, they've just been targeted with false information. A big part of what we're trying to do is correct that, number one.

Then, I have drafted five executive orders that are being looked at this week, that would resolve some of these problems. They're very specific areas that need to be addressed, and could only be done, I think, through an executive order from the president. That is something we're actively doing as of the past week. That's some of what's happening.

The other part that's very interesting concerns somebody who was with one of these national security agencies and connected to these old, illegal Legacy programs but would like a pathway out. Until there's an amnesty or safe harbor provision, people like him are terrified. Because they're obviously culpable for numerous crimes, up to and including treason, for which you can be executed.

This is a top priority, creating a glide path out of these clandestine programs that have been running illegally. Folks who now want to come forward but are in a world of trouble unless there's some kind of immunity or amnesty period. That is something that's being talked about very specifically. In fact, I saw some drafts of these executive orders that we recommended last year after the election that came out of the national security transition team.[423]

Heckenlively couldn't help but think that it always comes down to bureaucracy. People become part of a system; they realize that system isn't serving people and try to regain their humanity by telling the truth. The trouble is that the system may have already compromised them, even though now they want to do the right thing.

Michael had told Heckenlively that the new Trump administration had asked Greer to write a one-page briefing document to hand directly to the president. Heckenlively asked Greer about this claim. Greer confirmed that he had been asked to write such a document, had written it, and had been told it had been put in the president's hand. Heckenlively was able to obtain a copy of this document from an anonymous source. This is the text of it, with Greer's phone number and email redacted.

Memorandum

From: Steven M. Greer, MD—Director of the Disclosure Project
To: President Donald J. Trump
Re: The UFO/UAP subject

Since the 1950s, the UFO/UAP subject has been handled by a corrupt deep state transnational organization whose power has grown to a level that is an imminent threat to national security and international peace and security.

This organization is a hybrid of unconstitutional deep state and government compartmented operations and corporate special projects. It has reverse-engineered non-human intelligence (NHI) craft and is operating man-made advanced technologies at parity with NHI technologies. These human technologies are currently being used in a number of criminal operations including assassinations, abductions, human/drug/weapons trafficking, embezzlement of US government funds, acts of treason and have the capacity to simulate a fake alien attack at any moment.

I have debriefed over 700 government and corporate whistleblowers over the past 35 years and have documented their information in the Briefing Document provided to your staff.

The following Executive Orders are urgently needed.

1. Explicit whistleblower protection specifying both legal amnesty and personal security.

2. An Executive Order authorizing a TS-SCI SAP [top secret, secure communication infrastructure Special Access Program] with significant funding currently configured under law enforcement to stand down these illegal operations and especially the illegal use of electromagnetic pulse

weapons (EMP) currently being used against NHI craft as these actions imperil the future of humanity.

3. An Executive Order requiring all UFO/UAP operations to be fully disclosed within 6 months or those responsible will be vigorously prosecuted.

4. An Executive Order to authorize an advanced diplomatic team to make peaceful contact with NHI civilizations.

5. An Executive Order authorizing the review and release of Advanced Technology (AT) held by this criminal organization that would create total energy independence for the US and would begin a new energy economy with which the US would lead the world economically.

Please feel free to contact me at any time. I am the world's leading expert on this subject. There is not a distant second. I will provide any assistance, advice and evidence that you and your administration require.

Respectfully yours,
Steven M. Greer, MD
February 9, 2025[424]

According to Greer, this memorandum was put into Trump's hands, Trump read it, his eyes lit up, and he said, "I want to pursue this."[425]

And we think that's probably where all of us are at with this story. Many allegations have been made, and we can't tell you what is true. But we know that when a number of claims all point in a similar direction, it should engage your attention.

Even the skeptics we've interviewed for this book, such as Congressman Eric Burlison, believe that important secrets are being kept from the American public.

Perhaps we've accurately depicted what has been hidden for more than seventy years. Perhaps we've been fooled by an elaborate series of lies.

In the classic television series *The X-Files*, a common refrain is, "I want to believe." But it's not enough to believe or disbelieve in the alien phenomenon. Belief may be fine for matters of faith, such as whether or not there is a God who will judge our actions. However, in this world, there are things we can "know."

And more than anything else, we want to know whether aliens, non-human intelligences, unidentified flying objects, and/or unidentified aerial phenomenon are real, and whether a golden age of scientific miracles is at hand.

Can we cure disease, clean up our planet, feed the hungry, and journey to the stars?

We think it will be up to us to determine whether we stand at the brink of destruction or a new golden age, or if humans will just keep muddling blindly along in the madness of this nuclear age, perhaps suspecting much but knowing little. Greer is fond of claiming this is more of a spiritual question than anything else. And regardless of what one may believe about some of Greer's claims, he's proposed a clear path to resolving this question.

In the face of the unknown, do we choose hope over fear?

Do we choose courage over cowardice?

Perhaps we begin by refusing to accept the lies.

Perhaps it's by accepting that whatever the facts may be, we know that their full and complete disclosure will not be catastrophic.

We can handle the truth.

ENDNOTES

1   "UFO hearing key takeaways: What a whistleblower told Congress about UAPs," Stefan Becket, *CBS News*, updated on July 28, 2023, www.cbsnews.com/news/ufo-hearing-congress-uap-takeaways-whistleblower-conference-david-grusch-2023/.

2   "Unidentified Anomalous Phenomena: Implications on National Security, Public Safety, and Government Transparency," House Oversight and Accountability Subcommittee on National Security, the Border, and Foreign Affairs, effective July 26, 2023, www.congress.gov/event/118th-congress/house-event/116282/text.

3   Ibid.

4   Ibid.

5   Ibid.

6   Ibid.

7   Ibid.

8   Ibid.

9   Ibid.

10 Ibid.

11 Dr. Steven Greer, telephone interview by Kent Heckenlively, September 17, 2024.

12 "Dr. Steven Greer - Mystery Behind UFO / UAPs, Alien Phenomenon, and the Secret Government | SRS #048," *The Shawn Ryan Show*, YouTube, February 20, 2023, https://www.youtube.com/watch?v=NSLm__BUnmI.

13 Dr. Steven Greer, telephone interview by Kent Heckenlively, September 17, 2024.

14 Luis Elizondo, *Imminent: Inside the Pentagon's Hunt for UFOs*, (New York, New York: William Morrow, 2024), 2.

15 Ibid. at 82.

16 "Clapper Apologies for Answer on NSA's Data Collection," Bill Chappell, *NPR*, effective July 2, 2013, www.npr.org/sections/thetwo-way/2013/07/02/198118060/clapper-apologizes-for-answer-on-nsas-data-collection.

17 Luis Elizondo, *Imminent: Inside the Pentagon's Hunt for UFOs*, (New York, New York: William Morrow, 2024), 178.

18 Ibid. at 179.

19 David E. Sanger, *The Perfect Weapon: War, Sabotage, and Fear in the Cyber Age*, (New York: Crown Publishing Group, 2018), vii.

20 Luis Elizondo, *Imminent: Inside the Pentagon's Hunt for UFOs*, (New York, New York: William Morrow, 2024), 199.

21 Ibid. at 120.

22 Ibid. at 43.

23 Ibid. at 164.

24 Ibid. at 157–158.

25 Ibid. at 152.

26 Ibid. at 161–162.

27 Ibid. at 172.

28 Henry Luce, "The American Century," *Life*, February 17, 1941, 61, https://books.google.com/books?id=I0kEAAAAMBAJ&pg=PA61#v=onepage&q&f=false.

29 Ibid.

30 Ibid.

31 Ibid at 63.

32 Jacques F. Vallée and Paola Leopizzi Harris, *TRINITY: The Best Kept Secret*, (Starworks USA, LLC, 2021), 19.

33 Ibid. at 21–22.

34 Ibid. at 20.

35 Ibid. at 24.

36 Ibid. at 26–27.

37 Ibid. at 28.

38 Ibid. at 29.

39 Ibid. at 30–31.

40 Ibid. at 294.

41   Ibid. at 303–304.

42   Col. Phillip J. Corso (Ret.) with William Birnes, *The Day After Roswell*, (New York, New York: Gallery Books, 1997), 2–3.

43   "Senator Regrets Role in Book on Aliens," William J. Broad, *The New York Times*, effective June 5, 1997, https://www.nytimes.com/1997/06/05/us/senator-regrets-role-in-book-on-aliens.html.

44   Col. Phillip J. Corso (Ret.) with William Birnes, *The Day After Roswell*, (New York, New York: Gallery Books, 1997), 4.

45   Ibid. at 5.

46   Ibid. at 10–11.

47   Ibid. at 12.

48   Ibid. at 13–14.

49   Ibid. at 15.

50   Ibid. at 34–35.

51   Ibid. at 39.

52   Ibid. at 41.

53   Ibid. at 42–43.

54   Ibid. at 44–45.

55   Ibid. at 45.

56   Ibid. at 100.

57   Ibid. at 101–102.

58   Ibid. at 105.

59   Ibid. at 168.

60   Ibid. at 170.

61   Ibid. at 170–171.

62   Ibid. at 293.

63   Ibid. at 290.

64   Ibid. at 292–293.

65   Ibid. at 293.

66   Paola Leopizzi Harris, *Conversations with Colonel Corso*, (Boulder, Colorado: Luminous Moon Press, 2017) 29–30.

67   Ibid. at 73.

68   Ibid. at 35.

69   Ibid. at 36–38.

70 Ibid. at 97–98.

71 Ibid. at 42.

72 Ibid. at 32–33.

73 Paola Harris, interview by Kent Heckenlively and Michael Mazzola, November 11, 2023.

74 "The Top Secret Document Written by Oppenheimer and Einstein on Aliens and UFOs," Ancient Code Team, *Ancient Code*, effective March 15, 2025, https://www.ancient-code.com/the-top-secret-document-written-by-oppenheimer-and-einstein-on-aliens-and-ufos/.

75 "A Radar Blip, a Flash of Light: How U.F.O.s 'Exploded' into Public View," Laura M. Holson, *The New York Times*, effective August 3, 2018, https://www.nytimes.com/2018/08/03/science/UFO-sightings-USA.html.

76 Ibid.

77 "Albert Einstein's Letter About UFOs," Erin McCarthy, *Mental Floss*, effective January 18, 2023, www.mentalfloss.com/posts/albert-einstein-ufo-letter.

78 "Harry Truman Ordered this Alien Cover-Up," Mark Jacobson, *New York*, effective November 25, 2013, www.nymag.com/news/features/conspiracy-theories/harry-truman-aliens/.

79 Ibid.

80 Paola Harris, interview by Kent Heckenlively and Michael Mazzola, November 11, 2023.

81 Ibid.

82 Ibid.

83 "Is that JFK Memo to the CIA About UFOs real?" Natalie Wolchover, *NBC News*, April 21, 2011, www.nbcnews.com/id/wbna42704241.

84 Ibid.

85 Ibid.

86 Paola Harris, interview by Kent Heckenlively and Michael Mazzola, November 11, 2023.

87 Ibid.

88 Ibid.

89 Ibid.

90 Ibid.

91 Richard Dolan, *UFOs and the National Security State: Chronology of a Cover-up 1941–1973*, (Newburyport, Massachusetts: Hampton Roads Publishing, 2002), 35–36.

92 Ibid. at 36.

93 Christine LeFevre and Philip Lipson, *The Maury Island UFO Incident: The Story behind the Air Force's first military plane crash*, (Seattle, Washington: Northwest Museum of Lore and Legend, 2014), 12.

94 Richard Dolan, *UFOs and the National Security State: Chronology of a Cover-up 1941–1973*, (Newburyport, Massachusetts: Hampton Roads Publishing, 2002), 36.

95 Ibid. at p. 37.

96 Ibid.

97 Ibid.

98 Ibid.

99 Ibid. at 38.

100 Christine LeFevre and Philip Lipson, *The Maury Island UFO Incident: The Story behind the Air Force's first military plane crash*, (Seattle, Washington: Northwest Museum of Lore and Legend, 2014), 40.

101 Ibid. at 52.

102 Ibid. at 53.

103 Richard Dolan, *UFOs and the National Security State: Chronology of a Cover-up 1941-1973*, (Newburyport, Massachusetts: Hampton Roads Publishing, 2002), 39.

104 Ibid.

105 Christine LeFevre and Philip Lipson, *The Maury Island UFO Incident: The Story behind the Air Force's first military plane crash*, (Seattle, Washington: Northwest Museum of Lore and Legend, 2014), 63.

106 "Learn the Bizarre History of the Maury Island UFO Incident," Dan R. Green, *Medium*, effective November 7, 2022, https://medium.com/healthy-skepticism/learn-the-bizarre-history-of-the-maury-island-ufo-incident-5b28e138d654.

107 Christine LeFevre and Philip Lipson, *The Maury Island UFO Incident: The Story behind the Air Force's first military plane crash*, (Seattle, Washington: Northwest Museum of Lore and Legend, 2014), 63–64.

108 Ibid.

109 Ibid. at 65.

110 Ibid. at 67.

111 Ibid. at 68.

112 Ibid. at 69.

113 "Was Fred Crisman Behind the JFK Assassination?", Fred Litwin, *On the Trail of Delusion* updated October 9, 2021, www.onthetrailofdelusion. com/post/was-fred-crisman-one-of-jfk-s-assassins.

114 Ibid.

115 Kenneth Arnold and Ray Palmer, *The Coming of the Saucers*, (1952; reprinted, Zinc Read, 2023), 6.

116 Ibid. at 10.

117 Ibid.

118 Ibid. at 13.

119 Ibid.

120 Ibid. at 9.

121 Ibid at 14.

122 Ibid. at 17.

123 Ibid. at 33.

124 Ibid. at 33–34.

125 Ibid. at 34–35.

126 Ibid. at 37.

127 Ibid. at 42–43.

128 Ibid. at 42–43.

129 Kenneth Arnold, *Life: Report on Flying Saucers*, (Saucerian Publishers, 2021).

130 Ibid. at 6.

131 Ibid. at 8.

132 Ibid. at 16–17.

133 Ibid at 18.

134 Ibid.

135 Ibid.

136 Ibid. at 21–25.

137 Ibid. at 25.

138 Ibid.

139 Ibid. at 31.

140 "Washington's Blips: 'Somethings' over the capital are traced on radar," *Life*, August 4, 1952, 39, Google Books, https://books.google.com/books?id=YFYEAAAAMBAJ&pg=PA39#v=onepage&q&f=false.

141 Ibid. at 46.

142 Ibid. at 47.

143 Ibid, at 47–48.

144 "LIFE on the Newsfronts of the World," *Life*, August 11, 1952, 35, Google Books, https://books.google.com/books?idTFYEAAAAMBA&q=John+Samford#V=snippet&q=John%20Samford&f=false.

145 Albert K. Bender, *Flying Saucers and the Three Men*, (Clarksburg, West Virginia: Saucerian Books, 1962), 11.

146 Ibid. at 16.

147 Ibid. at 17.

148 Ibid. at 48.

149 Ibid. at 50–51.

150 Ibid. at 56–57.

151 Ibid. at 70.

152 Ibid. at 74.

153 Ibid. at 75.

154 Ibid. at 80.

155 Ibid. at 80–81.

156 Ibid. at 81.

157 Ibid.

158 John G. Fuller, *The Interrupted Journey*, (New York, New York: Vintage Books, 1966), xviii.

159 Ibid. at 5.

160 Ibid. at 19.

161 Ibid. at 20.

162 Ibid. at 23.

163 Ibid. at 65.

164 Ibid. at 166.

165 Ibid. at 120.

166 Ibid. at 122.

167 Ibid. at 127–129.

168 Ibid. at 130.

169 Ibid. at 144.

170 Ibid. at 168–169.

171 Ibid. at 246–247.

172 Ibid. at 248.

173 "Von Braun's Legacy of UFO and Space Tech | Dr. Carol Rosin Corporate Manager of Fairchild Industries," *Dr. Steven Greer*, YouTube, October 11, 2013, https://www.youtube.com/watch?v=gP8ftWzFYI4.

174 Ibid.

175 Ibid.

176 Ibid.

177 Ibid.

178 Ibid.

179 Ibid.

180 Ibid.

181 Ibid.

182 Ibid.

183 "Address to the 42nd Session of the United Nations General Assembly in New York, New York," Ronald Reagan Presidential Library and Museum, September 21, 1987, accessed August 20, 2023 www.reaganlibrary.gov/archives/speech/address-session-united-nations-general-assembly-new-york-new-york.

184 "Psychedelics: The newest tool in nuclear negotiations?", Robert K. Elder, *Bulletin of the Atomic Scientists*, effective December 17, 2021, www.thebulletin.org/2021/12/psychedelics-the-newest-tool-in-nuclear-negotiations/.

185 Ibid.

186 "US Space Force Chief of Space Operations General Raymond advocates dialogue with Russia on military space," Peace in Space, effective February 2, 2021, https://peaceinspace.com/2021/02/us-space-force-chief-of-space-operations-general-raymond-advocates-dialogue-with-russia-on-military-space/.

187 "Briefing Document: Operation Majestic 12: Prepared for President-Elect Dwight D. Eisenhower: (Eyes Only): November 18, 1952," Admiral Roscoe H. Hollenkoetter (MJ-1), Internet Archive, accessed September 4, 2024, https://ia800500.us.archive.org/35/items/majestic-12-documents-for-majic-eyes-only/Eisenhower%20Briefing%20Document_text.pdf.

188 "Project Grudge/Aquarius, Majestic 12 Group," The Majestic Documents, accessed September 4, 2024, www.majesticdocuments.com/investigations/official/project-grudge/.

189 "Flying saucer lands at Fort Eustis museum," Army Transportation Corps, accessed March 14, 2024, https://transportation.army.mil/history/studies/flying_saucer.html.

190 Timothy Good, *Earth: An Alien Enterprise,* (New York, New York: Pegasus Books, 2013), 45–46.

191 Ibid. at 47.

192 Ibid. at 49.

193 Ibid. at 49–50.

194 Ibid. at 54.

195 Ibid.

196 Ibid. at 54–55.

197 Ibid. at 55.

198 Ibid. at 56.

199 Ibid. at 59.

200 "Gordon Cooper's letter to Grenada's ambassador to the U.N." UFO Evidence, November 9, 1978, accessed September 4, 2024, http://www.ufoevidence.org/news/article161.htm.

201 Gordon Cooper with Bruce Henderson, *Leap of Faith: An Astronaut's Journey into the Unknown,* (New York, New York: Harper, 2000), 147.

202 Ibid. at 147–148.

203 Ibid at 148.

204 Ibid.

205 Ibid. at 184.

206 Ibid.

207 Ibid. at 184–185.

208 Harry A. Jordan, "USS Franklin D. Roosevelt—CVA-42—1962[–]1965," Disclosure Project of Dr. Steven Greer, www.majesticdocuments.com/investigations/witnesses-validating-majestic-documents/harry-a-jordan/.
209 Tony Craddock, email message to Dr. Steven Greer, February 24, 1999.
210 Paul Nahy, email message to Dr. Steven Greer, April 10, 2000, .
211 Ibid.
212 John W. Warner, IV to Dr. Steven Greer, October 27, 2016, in the recipient's possession.
213 Ibid.
214 Ibid.
215 A.J. Craddock, email message to Dr. Steven Greer, November 11, 1998.
216 "Report on the Historical Record of U.S. Governmental Involvement with Unidentified Anomalous Phenomena (UAP), Vol. I, February 2024," The Department of Defense All-domain Anomaly Resolution Office, cleared for open publication, March 6, 2024, https://www.aaro.mil/Portals/136/PDFs/AARO_Historical_Record_Report_Vol_1_2024.pdf.
217 Ibid. at 6.
218 Ibid.
219 Ibid. at 7.
220 Ibid.
221 Ibid.
222 Ibid. at 14.
223 Ibid.
224 Ibid. at 15.
225 Ibid.
226 Ibid. at 16.
227 Ibid.
228 Ibid. at 16–17
229 Ibid. at 17.
230 Ibid.
231 Ibid.
232 Ibid. at 18.

233 Ibid.

234 Ibid. at 19.

235 Ibid.

236 Ibid.

237 Ibid.

238 Ibid. at 20.

239 Michael Mazzola, email message to Kent Heckenlively, May 9, 2025.

240 "Report on the Historical Record of U.S. Governmental Involvement with Unidentified Anomalous Phenomena (UAP), Vol. I, February 2024," The Department of Defense All-domain Anomaly Resolution Office, cleared for open publication, March 6, 2024, 20, https://www.aaro.mil/Portals/136/PDFs/AARO_Historical_Record_Report_Vol_1_2024.pdf.

241 Ibid. at 21.

242 Ibid.

243 Ibid. at 21–22.

244 Ibid. at 22.

245 Ibid.

246 Ibid. at 22.

247 Ibid. at 23.

248 Ibid. at 24.

249 Ibid.

250 Ibid. at 24–25.

251 Ibid. at 25.

252 Ibid. at 25–26.

253 Ibid. at 29–30.

254 "EXCLUSIVE: Marine vet breaks 14-year silence to make astonishing claim that his six-man unit saw a hovering octagonal UFO being loaded with WEAPONS by unmarked US forces who threatened them at gunpoint while serving in Indonesia in 2023," Josh Boswell, *Daily Mail*, updated June 9, 2023, https://www.dailymail.co.uk/news/article-12177943/Marine-vet-breaks-14-year-silence-make-astonishing-claim-six-man-unit-saw-UFO.html.

255 Ibid.

256 Ibid.

257 Ibid.

258 "Report on the Historical Record of U.S. Governmental Involvement with Unidentified Anomalous Phenomena (UAP), Vol. I, February 2024," The Department of Defense All-domain Anomaly Resolution Office, cleared for open publication, March 6, 2024, 30, https://www.aaro.mil/Portals/136/PDFs/AARO_Historical_Record_Report_Vol_1_2024.pdf.

259 Ibid. at 31.

260 Ibid.

261 Ibid. at 32.

262 Ibid.

263 Ibid.

264 Ibid. at 33.

265 Ibid. at 34.

266 Ibid. at 34–35.

267 Ibid. at 40–45.

268 Email from Daniel Sheehan to Kent Heckenlively, June 15, 2025.

269 "Craft retrieval photos disprove AARO UAP report: Pentagon Papers lawyer | Reality Check," *NewsNation*, YouTube, March 11, 2024, https://www.youtube.com/watch?v=UVPs-2DfN_o.

270 "Daniel Sheehan - UFO Disclosure UPDATE – 2024," *Conscious Life Expo*, YouTube, December 22, 2024, https://www.youtube.com/watch?v=snFbna5Yftw.

271 Ibid.

272 Ibid.

273 Danny Sheehan, telephone interview by Kent Heckenlively, May 23, 2024.

274 Ibid.

275 Ibid.

276 Ibid.

277 Ibid.

278 Ibid.

279 Ibid.

280 Ibid.

281 Ibid.

282 Ibid.

283 Ibid.

284 Ibid.

285 "Some UFO Abductions Were Simulated CIA Psychological Warfare Experiments," Arjun Walia, *The Pulse*, effective March 23, 2025), https://www.thepulse.one/p/some-ufo-abductions-were-simulated.

286 Michael Mazzola, email message to Kent Heckenlively, May 9, 2025.

287 Jacques Vallée, *Forbidden Science 4: The Spring Hill Chronicles, The Journals of Jacques Vallée, 1990-1999*, (Anomalist Books, 2019), 114–115.

288 Danny Sheehan, telephone interview by Kent Heckenlively, May 23, 2024.

289 Ibid.

290 Ibid.

291 "Full special: Whistleblower reveals UAP retrieval program; object caught on video," Ross Coulthart, *NewsNation*, updated January 27, 2025, www.newsnationnow.com/space/ufo/hfr-uap-recovery-video-egg-shaped-object-exclusive/.

292 Danny Sheehan, telephone interview by Kent Heckenlively, May 23, 2024.

293 Congressman Eric Burlison, in person interview by Kent Heckenlively, July 12, 2024.

294 Ibid.

295 Ibid.

296 Ibid.

297 Ibid.

298 Ibid.

299 Ibid.

300 Ibid.

301 Ibid.

302 Ibid.

303 Ibid.

304 Ibid.

305 Ibid.

306 "House Hearing on Unidentified Anomalous Phenomena," Sub-committee on Cybersecurity, Information Technology, and Government Innovation; and Subcommittee on National Security, the Border, and Foreign Affairs, November 13, 2024, www.rev.com/transcripts/house-hearing-on-unidentified-anomalous-phenomena.

307 Ibid.

308 Ibid.

309 Ibid.

310 Ibid.

311 Ibid.

312 Ibid.

313 "Immaculate Constellation," Office of Congresswoman Nancy Mace, November 13, 2024, https://mace.house.gov/immaculateconstellation.

314 Ibid.

315 Ibid.

316 Ibid.

317 Ibid.

318 Ibid.

319 Ibid.

320 Ibid.

321 "House Hearing on Unidentified Anomalous Phenomena," Sub-committee on Cybersecurity, Information Technology, and Government Innovation; and Subcommittee on National Security, the Border, and Foreign Affairs, November 13, 2024, www.rev.com/transcripts/house-hearing-on-unidentified-anomalous-phenomena.

322 Michael Shellenberger, telephone interview by Kent Heckenlively, December 11, 2024.

323 Ibid.

324 Ibid.

325 Ibid.

326 "House Hearing on Unidentified Anomalous Phenomena," Sub-committee on Cybersecurity, Information Technology, and Government Innovation; and Subcommittee on National Security, the Border,

and Foreign Affairs, November 13, 2024, www.rev.com/transcripts/house-hearing-on-unidentified-anomalous-phenomena.

327 Ibid.

328 Congressman Eric Burlison, telephone interview by Kent Heckenlively, November 23, 2024.

329 Ibid.

330 Ibid.

331 Ibid.

332 "Mexican Congress Holds Second UFO Session Featuring Peruvian Mummies," Cassandra Garrison *Reuters*, updated November 7, 2023, https://www.reuters.com/world/americas/mexican-congress-holds-second-ufo-session-featuring-peruvian-mummies-2023-11-08/.

333 Ibid.

334 Michael Mazzola, telephone interview by Kent Heckenlively, January 17, 2024.

335 Ibid.

336 Ibid.

337 Ibid.

338 Ibid.

339 Michael Mazzola, telephone interview by Kent Heckenlively, February 12, 2024.

340 Ibid.

341 Ibid.

342 Ibid.

343 Ibid.

344 Ibid.

345 "DNA from 'non-human' alien corpses is from 'unknown species,' analysts claim," Melissa Koenig, *New York Post*, effective December 1, 2023, www.nypost.com/2023/12/01/news/dna-from-non-human-alien-corpses-is-from-unknown-species/.

346 Ibid.

347 "3-fingered 'alien mummies' found in Peru have fingerprints that do not appear to be human," Patrick Reilly, *New York Post*, effective July 23, 2024, https://nypost.com/2024/07/23/world-news/3-fingered-alien-

mummies-found-in-peru-have-fingerprints-that-do-not-appear-to-be-human-report/.

348 Ibid.

349 Ibid.

350 "Dr. John McDowell Named 2024 R.B.H. Gradwohl Laureate," *American Academy of Forensic Sciences*, effective November 15, 2023, www.aafs.org/article/dr-john-mcdowell-named-2024-rbh-gradwohl-laureate.

351 Joshua McDowell, telephone interview by Kent Heckenlively, July 29, 2024.

352 Ibid.

353 Ibid.

354 Ibid.

355 Ibid.

356 "Giant Mummies in Peru?," Joshua McDowell, *McDowell Law Blog*, May 3, 2024, www.mcdowellfirm.com/giant-mummies-in-peru/.

357 Joshua McDowell, telephone interview by Kent Heckenlively, July 29, 2024.

358 Ibid.

359 Ibid.

360 Ibid.

361 Ibid.

362 Ibid.

363 Dr. Jose Zalce, telephone interview by Kent Heckenlively, July 31, 2024.

364 Ibid.

365 "Doctor insists 'aliens are real' after examining corpses of strange creatures," Katherine Fidler, *Metro UK*, September 19, 2023, www.metro.co.uk/2023/09/19/aliens-mexico-real-fake-verdict-peru-19521013/.

366 Dr. Jose Zalce, telephone interview by Kent Heckenlively, July 31, 2024.

367 Ibid.

368 Ibid.

369 Ibid.

370 Ibid.

371 Ibid.

372 Ibid.

373 Ibid.

374 "Ricardo Martínez - Mexico," The Alien Project, accessed August 3, 2024, www.the-alien-project.com/en/mummies-of-nasca-mexico-2023/.

375 "Preliminary Report of DNA Study from Peruvian/Nazca Tridactyl Mummies," Dr. Ricardo Rangel Martínez, received by author on August 6, 2024.

376 Ibid.

377 Ibid.

378 Ibid.

379 Ibid.

380 Ibid.

381 Ibid.

382 Joshua McDowell, telephone interview by Kent Heckenlively, November 25, 2024.

383 Ibid.

384 Ibid.

385 Ibid.

386 Ibid.

387 Ibid.

388 Ibid.

389 Ibid.

390 Serena De Comarmond, telephone interview by Kent Heckenlively, November 27, 2024.

391 Ibid.

392 Ibid.

393 Ibid.

394 Ibid.

395 Ibid.

396 Edgar Hernàndez-Huaripaucar, Roger Zúñiga-Avilés, Bladmir Beccera-Cannales, Carlos Suarez-Canlla, Daniel Mendoza-Vizarreta, and Irvin Zúñiga-Almora, "Biometric Morpho-Anatomical Characterization and Dating of The Antiquity of A Tridactyl Humanoid Specimen: Regarding the Case of Nasca-Peru," *Revista De Gestão Social*

e Ambiental 18, no.5, May 28, 2024: www.doi.org/10.24857/rgsa. v18n5-137.

397 Text Message from Michael Mazzola to Eric Burlison and Kent Heckenlively, June 5, 2025, 5:42 PM.

398 Anthony Blair, "Scientists Discover Mysterious Sphere in Colombia, Sparking Speculation," *New York Post*, May 25, 2025, https://nypost. com/2025/05/25/science/scientists-discover-mysterious-sphere-in-colombia-sparking-ufo-speculation/.

399 Congressman Eric Burlison, telephone interview by Kent Heckenlively, June 27, 2025.

400 Ibid.

401 Ibid.

402 Ibid.

403 Mitchell Leslie, "Suddenly Smarter," *Stanford Magazine* (July/August 2002), www.stanfordmag.org/contents/suddenly-smarter.

404 Ibid.

405 Ibid.

406 Ibid.

407 "Origin of the Species, From an Alien View," Corey Kilgannon, *The New York Times*, effective January 8, 2010, https://www.nytimes. com/2010/01/10/nyregion/10alone.html.

408 Ibid.

409 Ibid.

410 Ibid.

411 Gen. 6:1–4, www.biblia.com/bible/niv/genesis/6/1-4.

412 Erich von Däniken, *Chariots of the Gods?*, (New York: G.P. Putnam & Sons, 1970), vii–ix.

413 "Catastrophic Disclosure - Episode 6," *Dr. Steven Greer*, YouTube, February 13, 2025 , https://www.youtube.com/watch?v=6Fr31_IOn9A.

414 Ibid.

415 Ibid.

416 Ibid.

417 Ibid.

418 Ibid.

419 Ibid.

420 Ibid.

421 Ibid.

422 Ibid.

423 Dr. Steven Greer, telephone interview by Kent Heckenlively, February 19, 2025.

424 Dr. Steven M. Greer, memorandum to President Donald J. Trump (provided anonymously), February 9, 2025.

425 Dr. Steven Greer, telephone interview by Kent Heckenlively, February 19, 2025.

ACKNOWLEDGMENTS

'd like to thank my lovely wife and partner in life, Linda; my amazing son Ben; and my daughter Jacqueline, who shows me the meaning of courage on a daily basis.

I'm appreciative of my father, who always supported my efforts, and my mother, who taught me that if a cause is just, you fight no matter the size of the enemy. I'd like to thank the best older brother in the world, Jay, who has been my champion and best friend since the day I was born. Joining the Heckenlively team is his amazing wife, Andrea, and their three wonderful children, Anna, John, and Laura. I'd be remiss if I didn't also mention the other up-and-coming writers in the family, Everette Glynn and Kate McLean. For somebody who's always considered himself "the kid" in the family, it's a strange experience to be treated as an "elder" by the younger members of my ever-expanding tribe. I will try to live up to your expectations of me.

. I'd like to thank some of the great teachers in my life: my seventh-grade science teacher, Paul Rago; high school teachers Ed Balsdon and Brother Richard Orona; college professors Clinton Bond, David Alvarez, and Carol Lashoff; writing teachers James Frey, Donna Levin, and James Alessandro; and, in law school, the always entertaining Bernie Segal.

I'd like to acknowledge the outside agitators who have tried five times to get me fired from my job as a middle school science teacher. It really is comical. As one administrator told me, "Kent, in the three years I've been here I don't think I've heard a single student or parent complaint about you. You are a revered teacher in this community." I like to think that's because I don't take myself too seriously, but I do take very seriously my responsibilities as a citizen of our great democracy and a public servant.

I've been blessed to have some of the greatest friends anybody could ever want, who have stood by me through success and failure. Thank you, John

Wible, John Henry, Chris Sweeney, Pete Klenow, Suzanne Golibart, Beth Bergen, Jyoti Dave, Sue Brown, Eric Holm, Sherilyn Todd, and Bill Wright.

I have a remarkable group of people whom I refer to as "Team Heckenlively," who allow me to keep up a ferocious writing schedule while also working as a science teacher and having a busy family life. There's my long-time editor, Max Swafford; my agent Johanna Maaghul; my fabulous publicist, Kelsey Merritt; my Muay Thai trainer, the amazing "Silky Smooth;" Eddie Abasolo, my coauthor on this book; Michael Mazzola; Henry Marx; Turu Marx; and the people who've championed my work in publishing, Caroline Russomano, Tony Lyons, Aleigha Koss, and Anthony Ziccardi.

Kent Heckenlively